Human Factors in Hazardous Situations

Human Factors in Hazardous Situations

Proceedings of
a Royal Society Discussion Meeting
held on 28 and 29 June 1989

Organized and edited by

D. E. Broadbent, F.R.S., J. Reason and A. Baddeley

CLARENDON PRESS · OXFORD
1990

Oxford University Press, Walton Street, Oxford OX2 6DP
Oxford New York Toronto
Delhi Bombay Calcutta Madras Karachi
Petaling Jaya Singapore Hong Kong Tokyo
Nairobi Dar es Salaam Cape Town
Melbourne Auckland
and associated companies in
Berlin Ibadan Nicosia

Oxford is a trade mark of Oxford University Press

Published in the United States
by Oxford University Press, New York

British Library Cataloguing in Publication Data

Library of Congress Cataloguing in Publication Data

ISBN 0 19 852191 X

First published in: Philosophical Transactions of the Royal Society of London, series B, Volume 327 (no. 1241), pages 447–593

Preface

Complex technology means that particular local failures can have widespread and severe consequences. This is true both in military and in civilian systems; chemical plants can create disasters just as nuclear weapons can. Thus we need to reduce all such failures to the lowest possible level. As engineering develops, there is a rise in the proportion of failures attributable to human error; this has produced a widespread interest in predicting and reducing such errors. This prompted a meeting at the Royal Society on 28 and 29 June 1989, to discuss aspects of the problem.

It is not possible for this volume to cover every aspect of this enormous field. For example, there is little discussion of the important problems of sociology and social psychology that arise in teams of controllers, there is little said about training and selection, and there are few comparisons between different methods of quantitative risk assessment. Instead, the book concentrates upon two main aspects of the problem.

One of these is the analysis of the decision process itself. The first three papers discuss different general aspects of such analysis, with a common conclusion that the nature of complex systems requires different methods of enquiry and explanation than those familiar from older technologies. The next three papers consider particular examples, the improvement of warning signals, the relation between actual performance and the individual's own account of it, and the findings obtained from investigating particular incidents in aviation.

The second theme is the impact on probability of failure, from background medical factors. The third group of papers take up some of these; the use of medication, the effects of time-of-day (shift work) or prolonged and sleepless activity, and the effects of infection. Finally, some examples of the possible approaches to practical action are touched upon; in decision aids, in analysis of industrial accidents, in a formal method of specifying thought processes, and in system design.

The original suggestion of a meeting on this topic came from Professor James Waterlow. He drew attention to the danger that the activities of researchers in particular laboratories were becoming isolated from the scientific community in general. We are grateful to his initiative, which we feel to have been justified by a valuable discussion; and we hope the volume also will help to break down such isolation.

December 1990

D. E. B.
A. D. B.
J. R.

Contents

PAGE

J. RASMUSSEN
Human error and the problem of causality in analysis of accidents [1]
Discussion: P. NIXON, F. WARNER [12]

D. DÖRNER
The logic of failure [15]
Discussion: P. NIXON, S. D. ROSEN [25]

J. REASON
The contribution of latent human failures to the breakdown of complex systems [27]

R. D. PATTERSON
Auditory warning sounds in the work environment [37]
Discussion: T. F. MAYFIELD [44]

D. E. BROADBENT, F.R.S.
Effective decisions and their verbal justification [45]

R. GREEN
Human error on the flight deck [55]

A. N. NICHOLSON
Medication and skilled work [65]

A. P. SMITH
Respiratory virus infections and performance [71]

M. F. ALLNUTT, D. R. HASLAM, M. H. REJMAN AND S. GREEN
Sustained performance and some effects on the design and operation of complex systems [81]

S. FOLKARD
Circadian performance rhythms: some practical and theoretical implications [95]
Discussion: S. D. ROSEN

J. FOX
Automating assistance for safety critical decisions [107]

PAGE

V. De Keyser
Temporal decision making in complex environments [121]

N. Moray
A lattice theory approach to the structure of mental models [129]

D. A. Norman
The 'problem' with automation: inappropriate feedback and interaction, not 'over-automation' [137]

Phil. Trans. R. Soc. Lond. B **327**, 447–593 (1990) [447]
Printed in Great Britain

HUMAN FACTORS IN HAZARDOUS SITUATIONS

A Discussion organized and edited by D. E. Broadbent, F.R.S., A. D. Baddeley and J. Reason

(*Discussion held* 28 *and* 29 *June* 1989 – *Typescripts received* 31 *July* 1989)

CONTENTS

PAGE

J. Rasmussen
Human error and the problem of causality in analysis of accidents 449
Discussion: P. Nixon, F. Warner, F.R.S. 460

D. Dörner
The logic of failure 463
Discussion: P. Nixon, S. D. Rosen 473

J. Reason
The contribution of latent human failures to the breakdown of complex systems 475

R. D. Patterson
Auditory warning sounds in the work environment 485
Discussion: T. F. Mayfield 492

D. E. Broadbent, F.R.S.
Effective decisions and their verbal justification 493

R. Green
Human error on the flight deck 503

A. N. Nicholson
Medication and skilled work 513

A. P. Smith
Respiratory virus infections and performance 519

M. F. Allnutt, D. R. Haslam, M. H. Rejman and S. Green
Sustained performance and some effects on the design and operation of complex systems 529

S. Folkard
Circadian performance rhythms: some practical and theoretical implications 543
Discussion: S. D. Rosen 553

J. Fox
Automating assistance for safety critical decisions 555

PAGE

V. De Keyser
Temporal decision making in complex environments 569

N. Moray
A lattice theory approach to the structure of mental models 577

D. A. Norman
The 'problem' with automation: inappropriate feedback and interaction, not 'over-automation' 585

Phil. Trans. R. Soc. Lond. B 327, 449–462 (1990)
Printed in Great Britain

Human error and the problem of causality in analysis of accidents

J. Rasmussen
Risø National Laboratory, DK 4000, Roskilde, Denmark

Present technology is characterized by complexity, rapid change and growing size of technical systems. This has caused increasing concern with the human involvement in system safety. Analyses of the major accidents during recent decades have concluded that human errors on part of operators, designers or managers have played a major role. There are, however, several basic problems in analysis of accidents and identification of human error. This paper addresses the nature of causal explanations and the ambiguity of the rules applied for identification of the events to include in analysis and for termination of the search for 'causes'. In addition, the concept of human error is analysed and its intimate relation with human adaptation and learning is discussed. It is concluded that identification of errors as a separate class of behaviour is becoming increasingly difficult in modern work environments. The influence of this change on the control of safety of large-scale industrial systems is discussed.

Introduction

During recent decades technological evolution has drastically changed the nature of the human factors problems one is faced with in industrial safety. In previous periods, the analysis of industrial systems could be reasonably well decomposed into problems that could be treated separately by different professions, and specialists in human factors were primarily involved in attempts to match the interface between people and equipment to human characteristics.

Complexity, rapid change and growing size of technical systems have drastically changed this state of affairs. The hazards involved in operation of large-scale systems lead to reliability and safety requirements that cannot be proven empirically. Consequently, design must be based on models that are able to predict the effects of technical faults and human errors during operation and to evaluate the ability of the operating organization to cope with such disturbances. The human factors problems of industrial safety in this situation not only includes the classical interface problems, but also problems such as the ability of the designers to predict and supply the means to control the relevant disturbances to an acceptable degree of completeness, the ability of the operating staff to cope with unforeseen and rare disturbances, and the ability of the organization in charge of an operation to maintain an acceptable quality of risk management. The human factors problems of industrial safety have become a true cross-disciplinary issue.

Analyses of certain major accidents during recent decades have concluded that human errors on part of operators, designers or managers have played a major role. However, to come from this conclusion to a suggestion of improvement is no easy matter and this problem appears to raise a couple of very basic issues related to the nature of causal explanations and to the concept of human error.

One basic issue to consider is the 'changing nature of engineering analysis', formerly a fairly well structured and bounded science. Technical designs could be verified and tested by quantitative models and controlled laboratory experiments. Descriptions in terms of mathematical relations among quantitative measurable variables were very effective for verification of design consistency and for optimization of the internal function, such internal functions as, for instance, the thermodynamic process of a steam locomotive. Description of the less well-formed relational structure of an accident such as a train crash was the business of another profession and the complexity of the situation required the use of less stringent causal analyses in terms of chains of events in which the internal functioning of the artefact is of minor importance; a locomotive in an accident is a heavy object, not a thermodynamical machine. This separation of the two domains of analysis is no longer acceptable for systems such as chemical plants, for which the internal functioning during accidents determines the external damage. The interaction of relational and causal analysis must therefore be better understood and the integration of the two methods of analysis must be improved.

Another issue is the notion of human error. The concept of error is challenged by the rapid technological evolution and the transfer of people from manual manufacturing tasks to supervision and intervention during disturbances in automated systems. In a traditional work setting, the slow pace of change led to the evolution of fairly stable work procedures and it was easy to define human error with reference to normal practice. Consequently, in traditional technology with slowly changing systems, causes of accidents were rather easy to determine and to attribute to technical faults or human error. Recently, however, some large-scale accidents point to the need for an explanation of accidents in terms of structural properties of integrated, large-scale systems rather than as isolated links or conditions in a linear causal chain of events.

Causal analysis of accidents

When analysing accidents after the fact, we are following a chain of events upstream to understand why it happened; to find somebody to blame, who done it; or find out how to improve the system. We are trying to describe a particular course of events and to identify the causes of the particular accident. It is, however, important to consider the implicit frame of reference of a causal analysis (Rasmussen 1988*a*).

Causal explanation

A classical engineering analysis is based on mathematical equations relating physical and measurable variables. The generalization depends on a selection of relations that are 'practically isolated' (Russell 1912). This is possible when they are isolated by nature (for example, being found in the planetary system) or because a system is designed so as to isolate the relation of interest (for example, in scientific experiment or a machine supporting a physical process in a controlled way). In this representation, material objects are only implicitly present in the parameter sets of the mathematical equations. The representation is particularly well suited for analysis of the optimal conditions and theoretical limits of physical processes in a technical system which, by its very design, carefully separates physical processes from the complexity of the outside world.

A causal representation is expressed in terms of regular causal connections of events. Russell (1912) discusses the ambiguity of the terms used to define causality: the necessary connection of events in time sequences. The concept of an 'event', for instance, is elusive: the more accurate the definition of an event, the lower the probability that it is ever repeated. Completeness removes regularity. The solution is, however, not to give up causal explanations. Representation of the behaviour of the physical world in causal terms is very effective for describing accidents because the objects of the real world are explicit in the model, and changes such as technical faults are easily modelled. This is not the case in a model based on relations among quantitative variables in which properties of an object are embedded in several parameters and equations. On the other hand, in seeking objective definitions, it must be realized that regularity in terms of causal relations is found between kinds of events, types, not between particular, individually defined events, tokens.

The behaviour of the complex, real world is a continuous, dynamic flow, which can only be explained in causal terms after its decomposition into discrete events. The concept of a causal interaction of events and objects depends on a categorization of human observations and experiences. Perception of occurrences as events in causal connection does not depend on categories that are defined by lists of objective attributes, but on categories that are identified by typical examples, prototypes (as defined by Rosch (1975)). This is the case for objects as well as for events. Everybody knows perfectly well what a 'cup' is. To define it objectively by a list of attributes that separates cups from jars, vases and bowls is no trivial problem and it has been met in many attempts to design computer programs for picture analysis. The problem is, that the property of being 'a cup' is not a feature of an isolated object but depends on the context of human needs and experience. The identification of events in the same way depends on the relation in which they appear in a causal statement. An objective definition, therefore, will be circular.

A classical example is 'the short-circuit caused the fire in the house' (Mackie 1965). This statement in fact only interrelates the two prototypes: the kind of short-circuit that can cause a fire in that kind of house. The explanation that the short-circuit caused a fire may be immediately accepted by an audience from a region where open wiring and wooden houses are commonplace, but not in a region where brick houses are the more usual kind. If not accepted, a search for more information is necessary. Short-circuits normally blow fuses, therefore further analysis of the conditions present in the electric circuit is necessary, together with more information on the path of the fire from the wiring to the house. A path of unusually inflammable material was probably present. In addition, an explanation of the short-circuit, its cause, may be needed. The explanation depends on a decomposition and a search for unusual conditions and events. The normal and usual conditions will be taken for granted, that is, implicit in the intuitive frame of reference. Therefore, in causal explanations, the level of decomposition needed to make it understood and accepted depends entirely on the intuitive background of the intended audience. If a causal statement is not accepted, formal logical analysis and deduction will not help, it will be easy to give counter-examples that cannot easily be falsified. Instead, further search and decomposition are necessary until a level is found where the prototypes and relations match intuition. (The reason that nuclear power opponents do not accept risk analysis may be that they have an intuition very different from a risk analyst's intuition, rather than a lack of understanding of risk and probability).

Accident analysis

The very nature of causal explanations shapes the analysis of accidents. Decomposition of the dynamic flow of changes will normally terminate when a sequence is found including events that match the prototypes familiar to the analyst. The resulting explanation will take for granted his frame of reference and in general, only what he finds to be unusual will be included: the less familiar the context, the more detailed the decomposition. By means of the analysis, a causal path is found upstream from the accident effect. This path will be prepared by resident conditions that are latent effects of earlier events or acts. These resident pathogens (Reason 1989) can themselves be explained by causal back-tracking and, in this case, branches in the path are found. To explain the accident, these branches are also traced backward until all conditions are explained by abnormal, but familiar events or acts. The point is: how does the degree of decomposition of the causal explanation and selection of the side-branches depend on the circumstances of the analysis? Another question is: what is the stop-rule applied for termination of the search for causes? Ambiguous and implicit stop rules will make the results of analyses very sensitive to the topics discussed in the professional community at any given time. There is a tendency to see what you expect to find; during one period, technical faults were in focus as causes of accidents, then human errors predominated, whereas in the future focus will probably move upstream to designers and managers. This points to the question whether system break-down is related to higher level functional structures and feed-back mechanisms rather than to the local conditions of events. In that case, traditional causal attributions turn out to be fighting symptoms rather than the structural origin of break-down.

The adoption of stop-rules is very important in the control of causal explanations. Every college student knows the relief felt when finding a list of solutions to math problems. Not that it gave the path to solution to any great extent, but it gave a clear stop-rule for the search for possible mistakes, overseen preconditions, and calculation errors. The result: hours saved and peace of mind. A more professional example to the same point is given by Kuhn (1976). He mentions the fact that chemical research was only able to come up with whole-number relations between elements of chemical substances after the acceptance of John Dalton's chemical atom theory. There had been no stop rule for the efforts in refinement of the experimental technique until the acceptance of this theory.

Stop-rules are not usually formulated explicitly. The search will typically be terminated pragmatically in one of the following ways: (a) an event will be accepted as a cause and the search terminated if the causal path can no longer be followed because information is missing; (b) a familiar, abnormal event is found to be a reasonable explanation; or (c) a cure is available. The dependence of the stop rule upon familiarity and the availability of a cure makes the judgement very dependent upon the role in which a judge finds himself. An operator, a supervisor, a designer, and a legal judge will reach different conclusions.

To summarize, identification of accident causes is controlled by pragmatic, subjective stop-rules. These rules depend on the aim of the analysis, that is, whether the aim is to explain the course of events, to allocate responsibility and blame, or to identify possible system improvements to avoid future accidents.

Analysis for explanation

In an analysis to explain an accident, the backtracking will be continued until a cause is found that is familiar to the analysts. If a technical component fails, a component fault will only be accepted as the prime cause if the failure of the particular type of component appears to be 'as usual'. Further search will probably be made, if the consequences of the fault make the designer's choice of component quality unreasonable, or if a reasonable operator could have terminated the effect, had he been more alert or better trained. In one case, a design or manufacturing error, in the other, an operator error will be accepted as an explanation.

In most recent reviews of larger industrial accidents, it has been found that human errors are playing an important role in the course of events. Very frequently, errors are attrubuted to operators involved in the dynamic flow of events. This can be an effect of the very nature of the causal explanation. Human error is, particularly at present, familiar to an analyst: to err is human, and the high skill of professional people normally depends on departure from normative procedures as we will see in a subsequent section. To work according to rules has been an effective replacement for formal strikes among civil servants.

Analysis for allocation of responsibility

To allocate responsibility, the stop-rule of the backward tracing of events will be to identify a person who made an error and at the same time, 'was in power of control' of his acts. The very nature of the causal explanation will focus attention on people directly and dynamically involved in the flow of abnormal events. This is unfortunate because they can very well be in a situation where they do not have the 'power of control'. Traditionally, a person is not considered in power of control when physically forced by another person or when subject to disorders such as, for example, epileptic attacks. In such cases, acts are involuntary (Fitzgerald 1961; Feinberg 1965), from a judgement based on physical or physiological factors. It is, however, a question as to whether cognitive, psychological factors should be taken more into account when judging 'power of control'. Inadequate response of operators to unfamiliar events depends very much on the conditioning taking place during normal work. This problem also raises the question of the nature of human error. The behaviour of operators is conditioned by the conscious decisions made by work planners or managers. They will very likely be more 'in power of control' than an operator in the dynamic flow of events. However, their decisions may not be considered during a causal analysis after an accident because they are 'normal events' which are not usually represented in an accident analysis. Furthermore, they can be missed in analysis because they are to be found in a conditioning side branch of the causal tree, not in the path involved in the dynamic flow.

Present technological development toward high hazard systems requires a very careful consideration by designers of the effects of 'human errors' which are commonplace in normal, daily activities, but unacceptable in large-scale systems. There is considerable danger that systematic traps can be arranged for people in the dynamic course of events. The present concept of 'power of control' should be reconsidered from a cognitive point of view, as should the ambiguity of stop-rules in causal analysis to avoid unfair causal attribution to the people involved in the dynamic chain of events.

Analysis for system improvements

Analysis for therapeutic purpose, that is, for system improvement, will require a different focus with respect to selection of the causal network and of the stop-rule. The stop-rule will now be related to the question of whether an effective cure is known. Frequently, cure will be associated with events perceived to be 'root causes'. In general, however, the effects of accidental courses of events can be avoided by breaking or blocking any link in the causal tree or its conditioning side branches. Explanatory descriptions of accidents are, as mentioned, focused on the unusual events. However, the path can also be broken by changing normal events and functions involved. The decomposition of the flow of events, therefore, should not focus on unusual events, but also include normal activities.

The aim is to find conditions sensitive to improvements. Improvements imply that some person in the system makes decisions differently in the future. How do we systematically identify persons and decisions in a (normal) situation when it would be psychologically feasible to ask for a change in behaviour as long as reports from accidents focus only on the flow of unusual events? An approach to such an analysis for improving work safety has been discussed elsewhere (Leplat & Rasmussen 1984).

Another basic difficulty is that this kind of analysis for improvement presupposes a stable causal structure of the system, it does not take into account closed loops of interaction among events and conditions at a higher level of individual and organizational adaptation. A new approach to generalization from analysis of the particular tokens of causal connections found in accident reports is necessary. The causal tree found by an accident analysis is only a record of one past case, not a model of the involved relational structure.

Human error, a stable category?

A number of problems are met when attempts are made to improve safety of socio-technical systems from analyses tied to particular paths of accidental events. This is due to the fact that each path is a particular token shaped by higher order relational structures. If changes are introduced to remove the conditions of a particular link in the chain, odds are that this particular situation will never occur again. We should be fighting types, not individual tokens (Reason & Wagenaar 1989). Human behaviour is constrained in a way that makes the chain of events reasonably predictable only in the immediate interface to the technical systems. The further away from the technical core, the greater the freedom agents have in their mode of behaviour. Consequently, the reference in terms of normal or proper behaviour for judging 'errors' is less certain. In this situation, improvements of safety features of a socio-technical system depend on a global and structural analysis: no longer can one assume the particular traces of human behaviour to be predictable. Tasks will be formed for the occasion, and design for improvements must be based on attempts to find means of control at higher levels than the level of particular task procedures. If, for instance, socio-technical systems have features of adaptation and self-organization, changes to improve safety at the individual task level can very well be compared to attempts to control the temperature in a room with a thermostat-controlled heater by opening the window. In other words, it is not sensible to try to change performance of a feedback system by changes inside the loop, mechanisms that are sensitive, i.e., related to the control reference itself, have to be identified.

In traditional, stable systems, human errors are related to features such as (*a*) conflicts among cognitive control structures and, (*b*) stochastic variability, both of which can be studied separately under laboratory conditions. In modern, flexible and rapidly changing work conditions and socio-technical systems, other features are equally important, such as: (*c*) resource limitations that turn up in unpredicted situations and finally, (*d*) the influence of human learning and adaptation. In the present context, the relation between learning and adaptation and the concept of error appears to be important.

Human adaptation

In all work situations constraints are found that must be respected to obtain satisfactory performance. There are, however, also many degrees of freedom which have to be resolved at the worker's discretion. In stable work conditions, know-how will develop, which represents prior decisions and choice, and the perceived degrees of freedom will ultimately be very limited, i.e., 'normal ways' of doing things will emerge, and the process of exploration necessary during the learning process will no longer be messing-up the concept of error. In contrast, in modern, flexible and dynamic work conditions, the immediate degrees of freedom will continue to require continuous resolution. This requires that effective work performance includes continuous exploration of the available degrees of freedom, together with effective strategies for making choice, in addition to the task of controlling the chosen path to a goal. Therefore, the basis of our concept of error is changed in a very fundamental way.

The behaviour in work of individuals (and, consequently, also of organizations) is, by definition, oriented towards the requirements of the work environment as perceived by the individual. Work requirements, what should be done, will normally be perceived in terms of control of the state of affairs in the work environment according to a goal, i.e., why it should be done. How these changes are made is to a certain degree a matter of discretion for the agent and cannot be predicted for a flexible environment.

The alternative, acceptable work activities, how to work, will be shaped by the work environment which defines the boundaries of the space of possibilities, i.e., acceptable work strategies. This space of possibilities will be further bounded by the resource profile of the particular agent in terms of tools available, knowledge (competence), information about state of affairs, and processing capacity. The presence of alternatives for action depends on a many-to-many mapping between means and ends present in the work situation as perceived by the individual; in general, several functions can serve each individual goal and each of the functions can be implemented by different tools and physical processes. If this was not the case, the work environment would be totally predetermined and there would be no need for human choice or decision.

Within the space of acceptable work performance found between the boundaries defined by the work requirements on the one hand, and the individual resource profile on the other hand, considerable degrees of freedom are still left for the individual to choose among strategies and to implement them in particular sequences of behaviour. These degrees of freedom must be eliminated by the final choice made by an agent to finally enter a particular course of action. The different ways to accomplish work can be categorized in terms of strategies, defined as types of behavioural sequences which are similar in some well defined aspects, such as the physical process applied in work and the related tools or, for mental strategies, the underlying kind of mental representation and the level of interpretation of perceived information. In actual performance, a particular situation-dependent exemplar of performance, a token, will emerge which is an implementation of the chosen strategy under the influence of the complexity of detail in the environment. The particular token of performance will be unique, impossible to predict, whereas the strategy chosen will, in principle, be predictable. This choice made by the individual agents depends on subjective performance criteria related to the process of work such as time spent, cognitive strain, joy, cost of failure, etc. Normally, dynamic shifting among alternative strategies is very important for skilled people as a means to resolve resource-demand conflicts met during performance.

Adaptation, self-organization and error

It follows directly from this discussion that structuring the work processes by an individual in a flexible environment will be a self-organizing, evolutionary process, simply because an optimizing search is the only way in which the large number of degrees of freedom in a complex situation can be resolved. The basic synchronization to the work requirements can either be based on procedures learned from an instructor or a more experienced colleague; or it can be planned by the individual on the specific occasion, in a knowledge-based mode of reasoning by means of mental experiments. From here, the smoothness and speed characterizing high professional skill, together with a large repertoire of heuristic know-how rules, will evolve through an adaptation process in which 'errors' are unavoidable side effects of the exploration of the boundaries of the envelope of acceptable performance. During this adaptation, performance will be optimized according to the individual's subjective 'process' criteria within the boundary of this individual resources. (That is, criteria based on factors applying during the work and not at the time of the ultimate outcome, or 'product'.) This complex adaptation of performance to work requirements, eliminating the necessity of continuous choice will result in stereotype practices depending on the individual performance criteria of the agents. These criteria will be significantly influenced by the social norms and culture of the group and organization. Very likely, conflict will be found between global work goals and the effect of local adaptation according to subjective process criteria. Unfortunately, the perception of process quality can be immediate and unconditional while the effect of the choice of a decision-maker on product quality can be considerably delayed, obscure and frequently conditional with respect to multiple other factors.

On the first encounter, if representation of work constraints is not present in the form of instructions from an experienced colleague or a teacher, and if know-how from previous experiences is not ready, the constraints of the work have to be explored in a knowledge-based mode from explicit consideration of the actual goal and a functional understanding of the relational structure of the work content. For such initial exploration as well as for problem solving during unusual task conditions, opportunity for test of hypotheses and trial-and-error learning is important. It is typically expected that qualified personnel such as process operators will and can test their diagnostic hypotheses conceptually, by thought experiments, before actual operations, if acts are likely to be irreversible and risky. This appears, however, to be an unrealistic assumption, since it may be tempting to test a hypothesis on the physical work environment itself in order to avoid the strain and unreliability related to unsupported reasoning in a complex causal net. For such a task, a designer is supplied with effective tools such as experimental set-ups, simulation programs and computational aids, whereas the operator has only his head and the plant itself. In the actual situation, no explicit stop rule exists to guide the termination of conceptual analysis and the start of action. This means that the definition of error, as seen from the situation of a decision maker, is very arbitrary. Acts that are quite rational and important during the search for information and test of hypothesis may appear to be unacceptable mistakes in hindsight, without access to the details of the situation.

Even if a human decision-maker is 'synchronized' to the basic requirements of work by effective procedures, there will be ample opportunities for refinement of such procedures. Development of expert know-how and rules-of-thumb depends on adaptation governed by subjective process criteria. Opportunities for experiments are necessary to find short-cuts and to identify convenient and reliable cues for action without analytical diagnosis. In other words, effective, professional performance depends on empirical correlation of cues to successful acts. Humans typically seek the way of least effort. Therefore, experts will not consult the complete set of defining attributes in a familiar situation. Instead it can be expected that no more information will be used than is necessary for discrimination among the perceived alternatives for action in the particular situation. This implies that the choice is 'under-specified' (Reason 1986) outside this situation. When situations change, for example, due to disturbances or faults in the system to be controlled, reliance on the usual cues that are no longer valid, will cause an error due to inappropriate 'expectations'. In this way, traps causing systematic mistakes can be designed into the system. Two types of error are related to this kind of adaptation: the effect of testing a hypothesis about a cue-action set, which turns out negative, and the effects of acts chosen from familiar and tested cues when a change in system conditions has made those cues unreliable.

Work according to instructions that take into consideration the possible presence of abnormal conditions that will make certain orders of actions unacceptable, presents an example in which local adaptation is very likely to be in conflict with delayed and conditional effects on the outcome. To be safe, the instruction may require a certain sequence of the necessary acts. If this prescribed order is in conflict with the decision-maker's immediate process criteria, modification of the prescribed procedure is very likely and will have no adverse effect in the daily routine. (If, for instance, a decision-maker has to move back and forth between several, distant locations because only that sequence is safe under certain infrequent, hazardous conditions, his process criterion may rapidly teach him to group actions at the same location together because this change in the procedure will not have any visible effect under normal circumstances.)

Even within an established, effective sequence of actions, evolution of patterns of movements will take place according to subconscious perception of certain process qualities. In a manual skill, fine-tuning depends upon a continuous updating of automated patterns of movement to the temporal and spatial features of the task environment. If the optimization criteria are speed and smoothness, adaptation can only be constrained by the once-in-a-while experience gained when crossing the tolerance limits, i.e., by the experience of errors or near-errors (speed-accuracy trade-off). Some errors, therefore, have a function in maintaining a skill at its proper level, and they cannot be considered a separate category of events in a causal chain because they are integral parts of a feed-back loop. Another effect of increasing skill is the evolution of increasingly long and complex patterns of movements which can run off without conscious control. During such lengthy automated patterns attention may be directed towards review of past experience or planning of future needs and performance becomes sensitive to interference, i.e., capture from very familiar cues.

The basic issue is that human errors cannot be removed in flexible or changing work environments by improved system design or better instruction, nor should they be. Instead, the ability to explore degrees of freedom should be supported and means for recovery from the effects of errors should be found.

System safety, adaptation, and error recovery

Dynamic adaptation to the immediate work requirements, both of the individual performance and of the allocation between individuals, can probably be combined with a very high degree of reliability; but only if errors are observable and reversible (i.e., critical aspects are visible without excessive delay), and individual process criteria are not overriding critical product criteria.

System breakdown and accidents are the reflections of loss of control of the work environment in some way or another. If the hypothesis is accepted that humans tend to resolve their degrees of freedom to get rid of choice and decision during normal work and that errors are a necessary part of this adaptation, the trick in design of reliable systems is to make sure that human actors maintain suffient flexibility to cope with system aberrations, i.e., not to constrain them by an inadequate rule system. In addition, it appears to be essential that actors maintain 'contact' with hazards in such a way that they will be familiar with the boundary to loss of control and will learn to recover (see the study of high reliable organizations described by Rochlin *et al.* (1989). In 'safe' systems in which the margins between normal operation and loss of control are made as wide as possible, the odds are that the actors will not be able to sense the boundaries and, frequently, the boundaries will then be more abrupt and irreversible. When radar was introduced to increase safety at sea, the result was not increased safety but more efficient transportation under bad weather conditions. Will anti-blocking car brakes increase safety or give more efficient transport together with more abrupt and irreversible boundaries to loss of control? A basic design question is: how can boundaries of acceptable performance be established that will give feedback to a learning mode in a reversible way, i.e., absorb violations in a mode of graceful degradation of the opportunity for recovery?

Under certain conditions self-organizing and adaptive features will necessarily lead to 'catastrophic' system behaviour unless certain organizational criteria are met. Adaptation will normally be governed by local criteria, related to an individual's perception of process qualities in order to resolve the perceived degrees of freedom in the immediate situation. Some critical product criteria (e.g., safety) are conditionally related to higher level combination or coincidence of effects of several activities, allocated to different agents and probably, in different time slots. The violation of such high level, conditional criteria cannot be monitored and detected at the local criterion level, and monitoring by their ultimate criterion effect will be too late and unacceptable. Catastrophic effects of adaptation can only be avoided if local activities are tightly monitored with reference to a prediction of their role in the ultimate, conditional effect, i.e., the boundaries at the local activities are necessarily defined by formal prescriptions, not active, functional conditions. (As argued below, the only possible source of this formal prescription is a quantitative risk analysis which, consequently, should be used as a risk management tool, not only as the basis for system acceptance.)

This feature of adaptation to local work requirements probably constitutes the fallacy of the defence-in-depth design principle normally applied in high-risk industries (Rasmussen 1988*b*). In systems designed according to this principle, an accident is dependent on simultaneous violation of several lines of defence: an operational disturbance (technical fault or operator error) must coincide with a latent faulty maintenance condition in protective systems, with inadequacies in protective barriers, with inadequate control of the location of people close to the installation etc. The activities threatening the various conditions normally belong to different branches of the organization. The presence of potential of a catastrophic combination of effects of local adaptation to performance criteria can only be detected at a level in the organization with the proper overview. However, at this level of the control hierarchy (organization), the required understanding of conditionally dangerous relations cannot be maintained through longer periods because the required functional and technical knowledge is foreign to the normal management tasks at this level.

The conclusion of this discussion is that catastrophic system breakdown is a normal feature of systems which have self-organizing features and at the same time, depend on protection against rare combination of conditions which are individually affected by adaptation. Safety of such systems depends on the introduction of locally visible boundaries of acceptable adaptation and introduction of related control mechanisms. What does this mean in terms of organizational structures? What kind of top-down influence from 'management culture' and bottom-up technological constraints can be used to guide and limit adaptation? How can we model and predict evolution of organizational structure and the influence on system safety?

Control of safety in high-hazard systems

The trend towards large-scale industrial process plants and the related defence-in-depth design practice point attention to the need for a better integration of the organization of plant design and of its operation. For large-scale, hazardous systems, the actual level of safety cannot be directly controlled from empirical evidence. For such installations, design cannot be based on experience gained from accidents, as has been the case for accidents in minor separate systems when, for instance, considering safety at work and in traffic. Consequently, the days of extensive pilot plant tests for demonstration of the feasibility of a design are also gone and safety targets have to be assessed by analytical means based on empirical data from incidents and near misses, i.e., data on individual, simple faults and errors. For this purpose, large efforts have been spent on developing methods for probabilistical risk analysis for industrial process plants.

Typically, however, such risk analysis is considered only for the initial acceptance of a particular plant design. It is generally not fully realized that a risk analysis is only a theoretical construct relating a plant model and a number of assumptions concerning its operation and maintenance to a risk figure. This fact implies that after the acceptance of a plant on the basis of the calculated risk, the model and assumptions underlying the risk analysis should be considered to be specifications of the preconditions for safe operation which, in turn, should be carefully monitored by the operating organization through the entire plant life (Rasmussen & Pedersen 1984).

This use of a risk analysis raises some important problems. Risk analysis and, in particular, the underlying hazard identification are at present an art rather than a systematic science. We have systematic methods for analysing specific accidental courses of events, the tokens. However, identification and characterization of the types of hazards to analyse, in particular related to the influence of human activities during operation, maintenance and plants management, to a large extent depend upon the creativity and intuition of the analyst as will be the case in any causal analysis. It is, therefore, difficult to make explicit the strategy used for hazard identification, the model of the system and its operating staff used for analysis, and the assumptions made regarding its operating conditions. Even if communication of causal arguments is unreliable between groups having different intuition such as designers and operations management, progress can be made, considering that the documentation of a risk analysis today is not designed for use during operations and maintenance planning and therefore is less accessible for practical operations management (Rasmussen 1988*b*).

Another problem is produced by the changing requirements of system management. Present organization structures and management strategies in industry still reflect a tradition that has evolved through a period when safety could be controlled directly and empirically. The new requirements for safety control based on risk analyses have not yet had the necessary influence on the predominant organizational philosophy. The basic problem is that empirical evidence from improved functionality and efficiency is likely to be direct and unconditional, when changes are made to meet economic pressure in a competitive environment. In contrast, decrease of safety margin in a 'safe' system caused by local sub-optimization, tends to be delayed and conditional and to require careful monitoring at higher management levels. Risk management requires a supplement of the traditional empirical managment strategies by analytical strategies based on technical understanding and formal analysis.

Conclusion

The conclusion of the arguments presented are that the present rapid trend towards very large and complex systems, process plants as well as systems for financial operations and for information storage and communication, calls for a reconsideration of some of the basic approaches to system design and operation and of the role of human error in system safety. Some of the deficiencies presently attributed to operator or management deficiencies may very well be structural problems which have to be considered at a much more fundamental level than efforts to improve human reliability.

References

Fitzgerald, P. J. 1961 Voluntary and involuntary acts. In *Oxford essays in jurisprudence* (ed. A. C. Guest). Oxford: Clarendon Press. Reprinted in *The philosophy of action* (ed. A. R. White). Oxford University Press.

Feinberg, F. 1965 Action and responsibility. In *Philosophy in America* (ed. M. Black). Allen and Unwin. Reprinted in *The philosophy of action* (ed. A. R. White). Oxford University Press.

Kuhn, T. 1962 *The structure of scientific revolution* University of Chicago Press.

Mackie, J. L. 1975 Causes and conditions. *Am. Phil. Q.* **2**, 245–255, 261–264. Reprinted in *Causation and conditionals* (ed. E. Sosa). Oxford University Press.

Rasmussen, J. & Leplat, J. 1984 Analysis of human errors in industrial incidents and accidents for improvement of work safety. *Accid. Anal. Prev.* **16**, 77–88.

Rasmussen, J. 1988*a* Coping safely with complex systems. *Am. Assoc. Ad. Sci.*, invited paper for Annual Meeting, Boston, February.

Rasmussen, J. 1988*b* Safety control and risk management: topics for cross-disciplinary research and development. Invited key note presentation in international conference on preventing major chemical and related accidents. In *IChemE Publication Series*, no. 110. Washington: Hemisphere Publishing.

Rasmussen, J. & Pedersen, O. M. 1984 Human factors in probabilistic risk analysis and in risk management. In *Operational safety of nuclear power plants*, vol. 1, pp. 181–194. Wien: IAEA.

Russell, B. 1913 On the notion of cause. *Proc. Aris. Soc.* **13**, 1–25.

Reason, J. 1990 *Human error: causes and consequences*. New York: Cambridge University Press.

Reason, J. 1986 Cognitive under-specification: its varieties and consequences. In *The psychology of error: a window on the mind* (ed. B. Baars). New York: Plenum Press.

Reason, J. & Wagenaar, W. 1990 Types and tokens in accident causation, CEC workshop. Workshop on Errors in Operation of Transport Systems; MRC-Applied Psychology Unit, Cambridge.

Rosch, E. 1975 Human categorization. In *Advances in cross-cultural psychology* (ed. N. Warren). New York: Halstead Press.

Rochlin, G. I., La Porte, T. R. & Roberts, K. H. 1987 The self-designing high-reliability organization: aircraft carrier flight operations at sea, *Naval War Coll. Rev.* **40**, 76–90.

Discussion

P. Nixon (*Charing Cross Hospital, London, U.K.*). The concept of 'boundaries' of tolerance is important in medicine because people push themselves or allow themselves to be driven beyond their physiological territory into boundary-testing. Some live with sick or inadequate systems for years on end without having the energy, the information, or the opportunity for recovery. The label 'MD' or 'post-viral syndrome' does not point out the remedy. Others hurl themselves into the catastrophe of a heart attack without doctors teaching them to recognize their tolerance.

J. RASMUSSEN. Thank you for the reference to the medical scene. Also in this field, the problems involved in using causal analysis for therapeutic purposes are being discussed, and several fundamental parallels have been found when analysing medical cases and records for accidents in large-scale complex socio-technical systems. In both cases, the problem has been met to identify, from analysis of particular cases, the influence of inherent functional cross-couplings and adaptive features.

F. WARNER, F.R.S. (*University of Essex, U.K.*). The original title of the meeting (Human factors in high-risk situations) should be queried before discussion begins, and regard paid to the definitions in the Royal Society Report 'Risk Assessment'. The situations discussed may represent major hazards but they are low risk. The Report reminds us that the high risks are met in normal life, of a little more than 10^{-2} per year for life expectancy but doubled for a President of the U.S.A.; around 10^{-4} per year for motor accidents in cars in the U.K., but 10^{-2} per year for motor cycles driven by the 17–24-year age group (against a risk of 10^{-4} per year of death in this group). People voluntarily accept higher risks in hang-gliding or rock climbing but have worried recently about deaths in rail accidents where the risk is just over 3×10^{-6} per year, or about aircraft where it is half this level. The figures are, of course, for deaths that are certain and available.

There is another way in which risk can be high and related to financial loss. The detailed information is retained by insurers for whom it is part of their stock in trade. Some approximate assessment can be made by comparing insurance premiums. It is difficult to find any smaller than 10^{-3} per year for buildings, household property and car insurance, the familiar areas. It will be interesting to learn from this meeting the levels in aviation insurance where underwriters normally reckon business to be profitable. They would show that the financial risks are lower, if not as low as those for death.

The paper queries models that are quantitative and deterministic but hazard and operability studies are probabilistic and failure rates allow for this. The process of analysis does allow for critical examination of assumptions made from the supplier and customer standpoint with independent audit. The aircraft industry had to take the lead in this because deterministic methods could not produce viable answers. Its influence has spread to quality assurance and the methodology of BS 5750/ISO 9000 applicable not only to products but services and systems. The application of this requires human participation and understanding of the operation at the level of those taking part.

J. RASMUSSEN. Sir Frederick raises two points which I probably should have been more clear about in my talk: First, I whole-heartedly agree that in most cases with well planned, modern activities, high-hazard low-risk situations rather than just high-risk situations are in focus. This is the case even in major chemical installations and nuclear power systems. This issue has been discussed also in the current series of workshops sponsored by the UN World Bank on organizational issues for risk management in high-hazard low-risk systems. The other issue raised by Sir Frederick is the question of probabilistic risk analysis. I apologise if my presentation leads one to believe that I am criticising predictive, probabilistic risk analysis as a conceptual tool. Far from that, I believe that such quantitative prediction is the only way towards an acceptable risk management in high-hazard systems. What I am criticizing is the present use of risk analysis as mainly being a tool for acceptance of new installations. Instead it should be the most important tool for operations management, which however, requires considerable further methodological development to make the preconditions of risk analysis explicit and to communicate them to operations management in an effective way. Major accidents in the past have typically been released by a systematic erosion of the preconditions of the quantitative predictions, not by a stochastic coincidence of independent events. This makes the prediction unreliable but does not affect the value of predictive as a tool for identification of the necessary preconditions to monitor during plant life so as to operate at the predicted level of safety.

Phil. Trans. R. Soc. Lond. B **327**, 463–473 (1990)
Printed in Great Britain

The logic of failure

BY D. DÖRNER
Universität Bamberg, Lehrstuhl Psychologie II, Bamberg D-8600, F.R.G.

Unlike other living creatures, humans can adapt to uncertainty. They can form hypotheses about situations marked by uncertainty and can anticipate their actions by planning. They can expect the unexpected and take precautions against it.

In numerous experiments, we have investigated the manner in which humans deal with these demands. In these experiments, we used computer simulated scenarios representing, for example, a small town, ecological or economic systems or political systems such as a Third World country. Within these computer-simulated scenarios, the subjects had to look for information, plan actions, form hypotheses, etc.

1. Introduction

(a) *A famine in West Africa*

Our World, unfortunately, is not exactly bereft of catastrophes. Three years ago, the explosion of the Chernobyl reactor made for a great deal of anxiety and a great many discussions, which are still going on, regarding the extent to which humanity can afford technology, the failure of which has such far-reaching consequences.

Questions like these are, of course, especially salient when, as in the case with the Chernobyl disaster, one is dealing with 'man made' catastrophes rather than natural ones. This case, after all, poses the question of avoidability in an especially urgent manner.

The catastrophe I shall be concerned with now, occurred in West Africa, in the Moro region in the state of Bukina Faso which is shown in figure 1. That catastrophe hardly made any headlines. Twelve years ago a development aid project was begun there with the aim of creating generally better living conditions for the semi-nomadic people living in this region. By sinking deep wells, the available water supply was increased, the pasture areas for cattle were better irrigated, resulting in a sharp increase in the stock of herds of cattle. In addition, the tsetse fly, which had severely infected the cattle with sleeping sickness, had been brought under control. So the Moros ten years later were considerably better off. The development of some important variables of the Moro region can be seen in figure 2.

In the end, herds of cattle exceeded the capacity of the available pasture area; a slight decrease in rainfall led to a food shortage; the pastures were overgrazed, and the hungry cattle destroyed the turf and hence the basis for vegetation: the cattle starved to death, thus giving rise to a famine in the human population, which claimed many lives.

Fortunately, this catastrophe did not actually take place in West Africa, but in an office in downtown Zurich. The cause of the catastrophe was not due to a team of development aid volunteers but an executive of an international computer enterprise. The Moro region merely existed in the form of a computer program and serves us as an instrument for investigating the action strategies implemented by persons in coping with extremely complex, dynamic, intransparent and uncertain systems. I shall now describe these attributes of a system more closely.

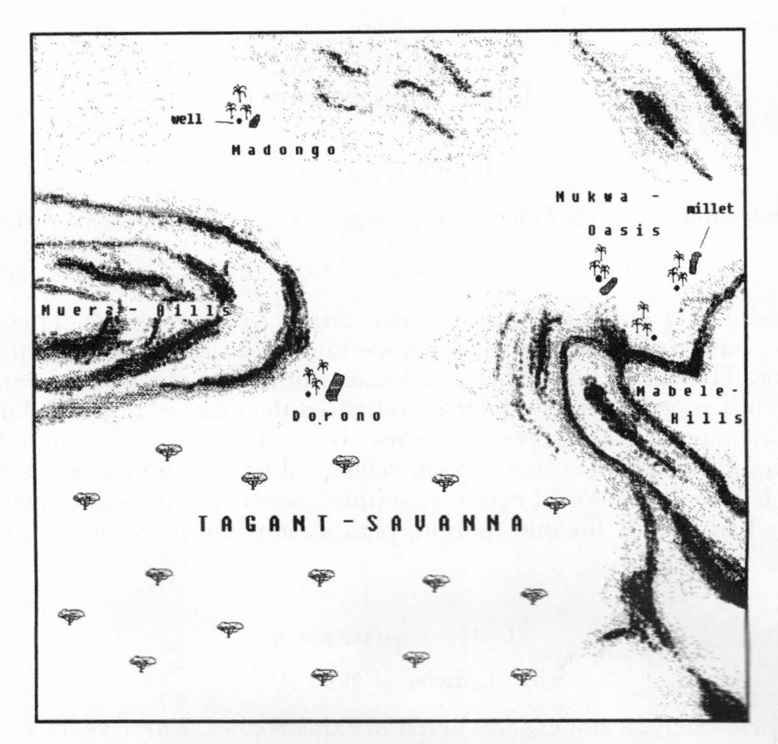

FIGURE 1. The Moro region in Bukina Faso, West Africa.

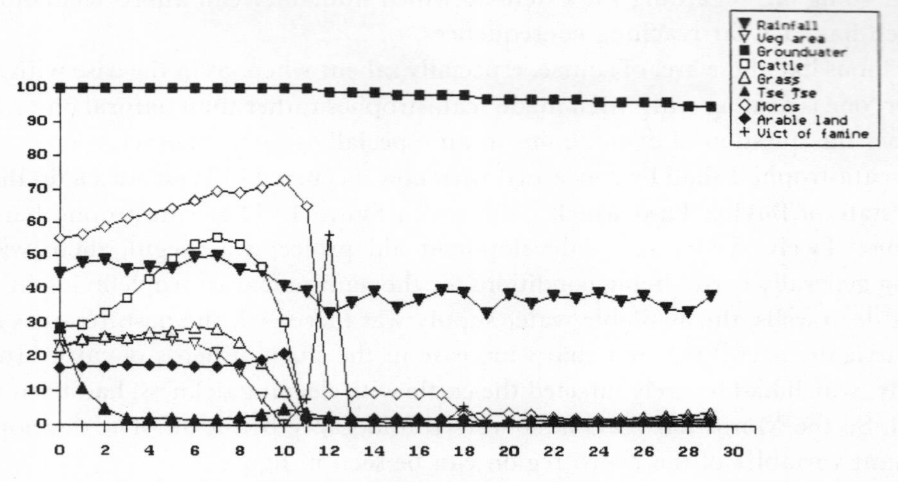

FIGURE 2. A famine in the Moro region.

A system is extremely complex when it consists of a great variety of variables. The Moro system consists of the following variables: population with birth-rate and mortality, cattle stock, ground water, vegetation area, precipitation, area of arable land, millet harvest, etc. All these variables are closely tied to one another, they mutually affect each other and constitute a network of interdependencies, which are shown in figure 3.

The Moro system is dynamic, which means that it develops further, even without interventions. It does not remain stable and waits for interventions like, for instance, a chess board. Because the Moro system is intransparent, many of the variables, for example, the

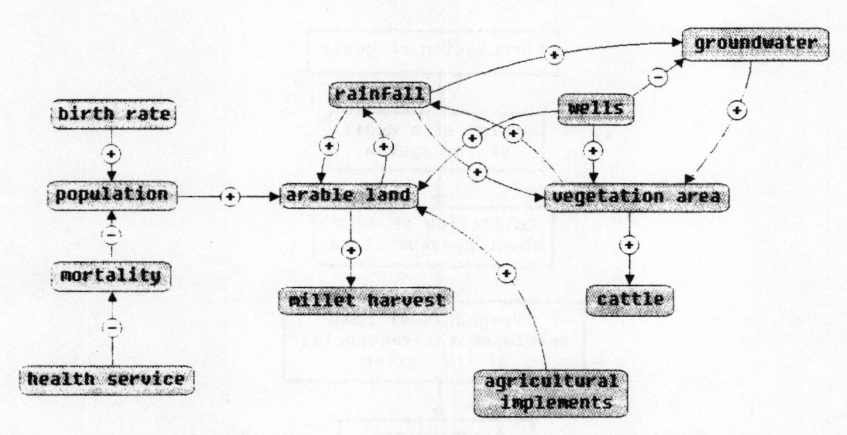

FIGURE 3. Part of the network structure of the Moro system.

groundwater level, defy direct observation. The uncertainty of the Moro system means that the acting subject has no complete knowledge about the system, about its variables and their interdependencies.

For our experimental subjects, the Moro system is a 'dynamic decision problem' (Brehmer & Allard 1986) as the experimental subjects are included in the development of the system. They gather information at a certain point in time, make decisions for particular measures and then execute these measures; they are confronted with the effects of their decisions, are able to gather further data, decide for the same or new measures, and so on. Figure 4 shows the results of the governmental activities of a good subject. It is possible to create positive living conditions for the Moros over a long period of time.

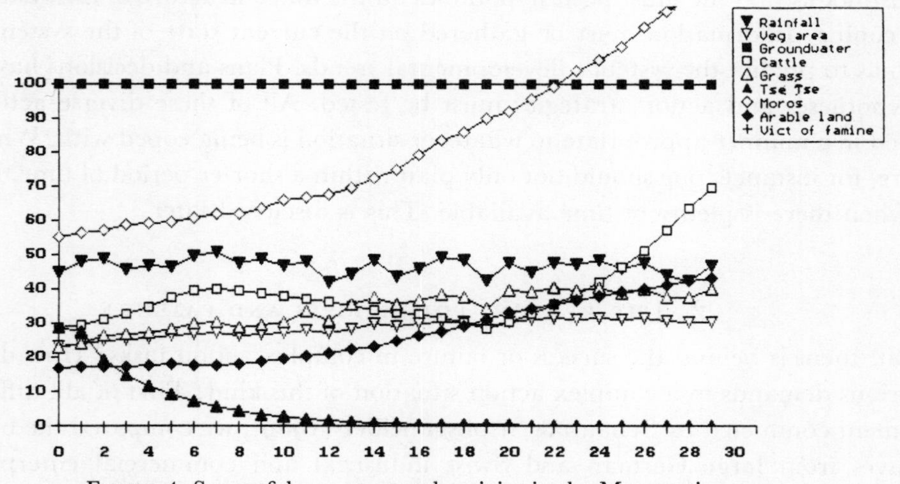

FIGURE 4. Successful governmental activity in the Moro region.

Of course, we do not investigate the behaviour of our experimental subjects in the Moro region with the intention of learning something about the action strategies of development aid volunteers in West Africa. Instead, we believe that action situations, such as those in the Moro region, are in many respects typical of the conditions for action in the rest of the world.

The reason for our experiments with computer-simulated environments is to gain a general insight into the psychology of action-regulation in complex fields of reality. People have to meet

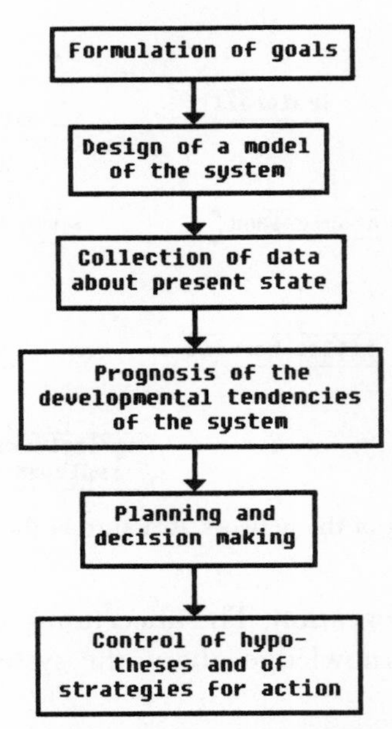

FIGURE 5. Demands of complex action regulation.

various demands in such a situation, which are not at all typical only for the Moro system. Figure 5 shows the tasks that must be performed in coping with such a system.

The first important task is that of formulating goals. More often than not, one has to first get the goals clear in one's mind, as the goals in complex action situations are often given in an indistinct form. One must posit hypotheses on the inner structure of the system with which one is coping. Information must be gathered on the current state of the system at any given time so as to predict the system's developmental trends. Plans and decisions have to be made, and hypotheses and action strategies must be tested. All of these diverse activities must be executed in a manner appropriate to whatever situation is being coped with. When under time pressure, for instance, one should not only plan within a shorter period of time, but differently than when there is plenty of time available. This is discussed later.

2. The reasons for success and failure

What, then, is behind the success or failure in complex action tasks? How do people meet the various demands in a complex action situation of this kind? First of all, differently! In an experiment conducted by Schaub & Strohschneider (1989), we compared the behaviour of 45 executives from large German and Swiss industrial and commercial enterprises with the behaviour of students. Each subject had to try, over a (simulated) period of over 20 years, to improve the living conditions of the Moros. In each year, the subjects were able to ask questions about the (then) current state of the Moros and to make any decisions they wished to. They could buy tractors for the Moros, sink wells, improve medical care, influence trade relations and so on. Figure 6 shows the average results for the executives and the students. On the whole, the executives did far better; they earned, by skilful management, more capital, their herds of cattle were greater, as were their vegetation areas and human populations. There were, accordingly, fewer catastrophes in their simulations.

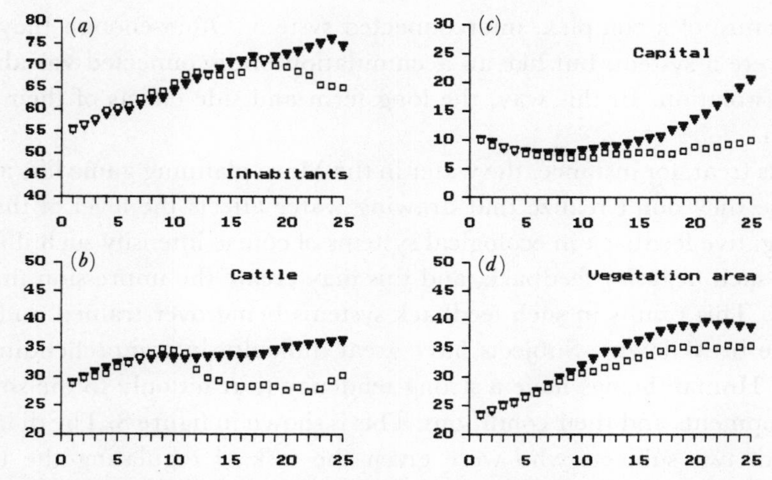

FIGURE 6. The average developments in (*a*), population; (*b*), cattle; (*c*), capital; and (*d*), vegetation, for executives (triangles) and students (squares).

Which mistakes do people make when faced with coordinating complex actions? What are the reasons for catastrophes like the Moro catastrophe described above? We found a great variety of modes of faulty behaviour that were responsible for the failure when acting in such complex situations. Figure 7 presents a rough overview of the various forms of faulty behaviour. Subjects fail, for instance, to get their goals clear and then act according to a 'repair service policy'. They do this by eliminating the obvious errors and solving the conspicuous problems, while disregarding the less conspicuous ones and, of course, failing to take into account aberrant developments that first become apparent in faint symptoms. Subjects fail to construct

insufficient goal elaboration
→ acting according to a 'repair-shop' principle

insufficient formation of hypotheses about the structure of the system
→ neglecting side- and long-term effects

insufficient ideas about the behaviour of the system in time
→ neglecting developmental tendencies of the system

insufficient coordination of different measures
→ 'collision' of measures

'ballistic' action
→ no detection of wrong hypotheses and inappropriate strategies

no self-reflection
→ no 'repair' of wrong hypotheses and inappropriate strategies

FIGURE 7. Modes of faulty behaviour in coping with complex systems.

an adequate picture of a complex, interconnected system. Often enough they do not treat a system as if it were a system, but like an accumulation of disconnected variables that can be manipulated in isolation. In this way, the long term and side effects of their actions remain obscure to them.

Many subjects treat, for instance, the water in the Moro planning game like an inexhaustible resource because they don't realize that drawing water affects the level of the ground water (instances of negative feedback in ecological systems of course intensify such illusions. Stability is generated by such negative feedback, and this may create the impression that a system can 'take anything'. This results in such feedback systems being overstrained and, ultimately, in their irreversible breakdown). Subjects have great difficulty in comprehending the temporal forms of events. Human beings have a strong tendency to react only to the status quo and to disregard developments and their conditions. This is shown in figure 8. The diagram represents the behaviour of two subjects who were given the task of regulating the thermostat of a temperature control unit such that the temperature of 4 °C was attained in a cool-storage unit. The thermostat was a time-delayed system, hence a system with damped sine oscillations, and the subjects had to discover the correct dial setting. As you can see, the subject shown in figure 8 (*b*) managed this rather well, while the subject shown in 8 (*a*) had more difficulty regulating the system. This latter subject reacted in each case to the immediate temperature state, failed to include the oscillations of the system in the calculations, never learned to do so and therefore always met with failure in regulating the thermostat. (See Reichert & Dörner 1988.)

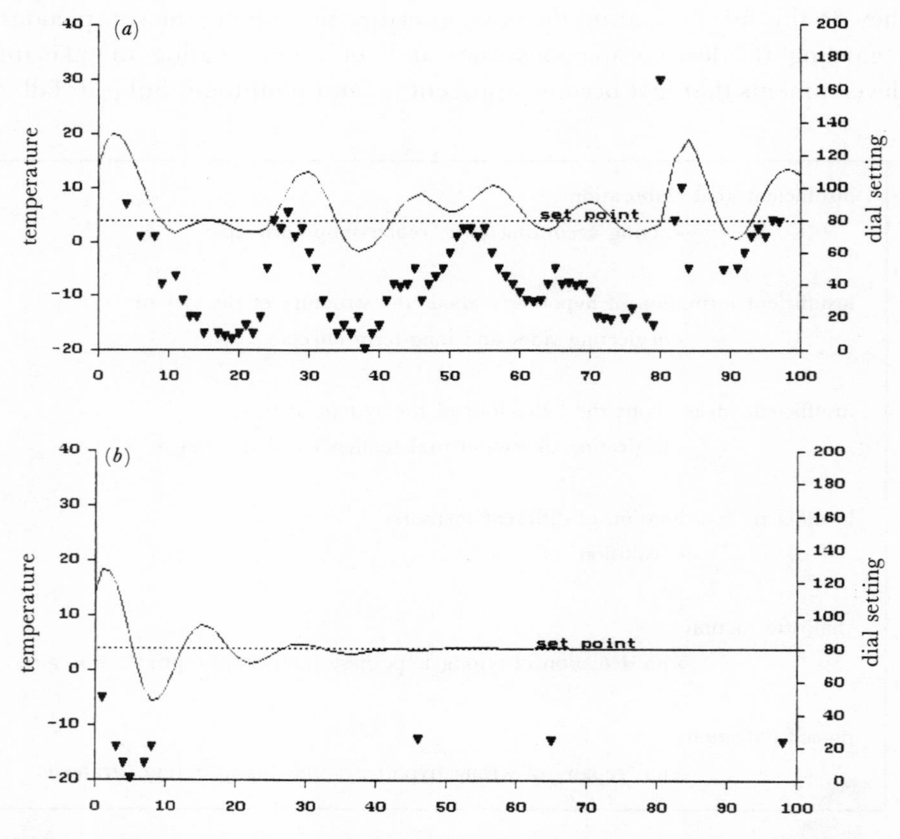

FIGURE 8. (*a*) 'Doer' behaviour and (*b*) 'wait-and-see' behaviour in regulating a thermostat.

Other experiments yielded similar results. The transition from a mode of behaviour focused merely on the momentary state to one which includes the dynamics of a system appears to be extremely difficult. Subjects prognosticate the development of variables by means of linear extrapolation. They assume a linear development even when the situation clearly shows that the development must be nonlinear. They fail to take into account nonlinear developments and especially sudden, 'catastrophic' developments. At the same time, it seems difficult to coordinate various lines of action. Figure 9 shows the successful and unsuccessful subjects in the Moro experiment. You can see that for the unsuccessful subjects the questions and decisions are mixed. This means that the subjects made immediate decisions for each of the states they inquired about. For the successful subjects, however, the question and decision behaviour was not mixed. This means that the successful subjects first gained an overall picture of all aspects of the systems before they began to make their first decisions.

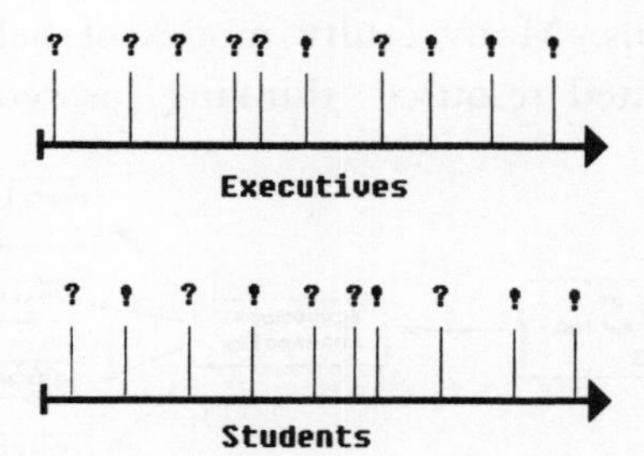

FIGURE 9. Scheme of the sequencing of questions and measures for successful and unsuccessful subjects.

Subjects often act 'ballistically'. They take measures without checking the effects of these measures later. As the effects of measures are usually uncertain, this lies in the nature of complex systems, this is a dangerous error. Crisis situations are especially susceptible for ballistic forms of action, as shown in figure 10. Here, the percentage of checked measures is depicted for an experiment of the Moro type, which was conducted by Reither (1985). You can see that in the first phases of the experiment the experimental subjects checked only between 30 to 50% of their measures. After a 'crisis' in the 10th year, which consisted of an unexpected military aggression of a neighbouring state, the percentage of checked measures

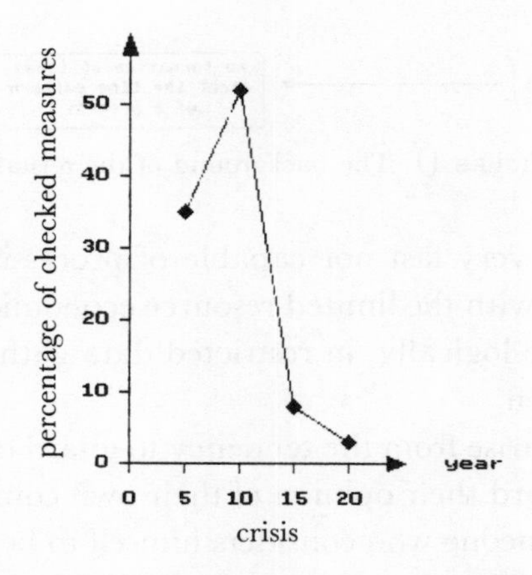

FIGURE 10. Percentage of checked measures in the Reither experiment.

decreased to below 10%. The experimental subjects seemed to exhibit a tendency not to be confronted with the effects of their measures so as to be able to maintain an illusion of competence. Subjects dispense with the self-reflexive analysis of their own behaviour. Subjects' strategies for coping with complex systems are for the most part insufficient, in one respect or another. Self-reflexive examination and critique of one's own way of acting is an essential means of adapting one's own way of acting to the given circumstances. Dispensing with self-reflection is therefore a major error.

3. The background of the mistakes

What is the psychological background of the mistakes just described? If one tries to answer this question, it must be made clear that, while the faulty modes of behaviour described above are inadequate for coping with a complex system, they serve other purposes rather well. Figure 11 shows some interconnections. Many faulty modes of behaviour are expressions of the tendency to deal with the limited resource 'thinking' as economically as possible. Human

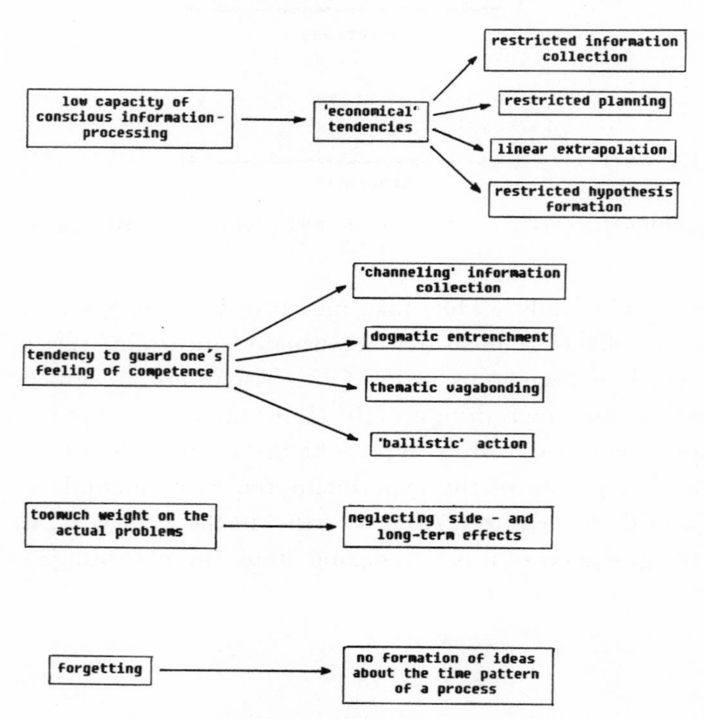

Figure 11. The background of the mistakes.

conscious thinking is neither very fast nor capable of processing very much information per time unit. A tendency to deal with the limited resource economically may therefore make sense. This results, in a certain sense logically, in restricted data gathering, linear extrapolation and restricted hypothesis formation.

Other modes of behaviour arise from the tendency to guard one's own competence. Humans have a strong tendency to guard their opinion of their own competence in acting. To a certain extent this makes sense, as someone who considers himself to be incapable of acting will hardly act. Guarding one's opinion of one's competence is an important motivation. But it can lead to deformations in the thought process. To maintain a high opinion of one's own competence, people fail to take notice of data that show that their hypotheses are wrong. Or they act 'ballistically' and do not check the effects of their actions so as to maintain the illusion of having solved the corresponding problems by means of their action. The underlying reasons for dispensing with self-reflection may also lie in the tendency to avoid looking at one's own mistakes so as not to endanger one's estimation of one's own competence.

Still other modes of behaviour come from the 'predominance of current problems'. Subjects deal with the problems they have, not with those they do not have. Why is this wrong? Why should one bother about problems one does not have? In dynamic systems, the problems one does not have are usually the problems one does not yet have. One's whole attention should not be devoted to the current problems; one must consider possible developments and the possible long term and side effects of the measures taken.

Adapting oneself inadequately to the sequential characteristics of processes may also be attributable to an incredibly simple feature of human data processing, namely, forgetfulness. An important requirement for gaining the correct picture of temporal sequences is having information on the length of time available. If this is not the case, one is also unable to posit hypotheses on temporal patterns. The fact that people forget means that past data are only partially available. This means that there are great difficulties in recognizing the correct form of temporal sequences. A simple means of coping with this difficulty is the 'spatialization' of time. Diagrams of temporal sequences make it possible to treat temporal sequences like 'spatial forms', which are easier to cope with.

4. Strategic flexibility

Why are the executives tested in the Schaub–Strohschneider study less subject to the effects of the factors just stated than the students? There are clues that show that this is because the executives know better 'when to do what'. To a greater extent than the students, the executives have at their disposal a system of local rules (that is rules to be deployed only under certain conditions) for specifically coping with problem situations. And that system of local rules gives them strategic flexibility. Let us for example consider planning behaviour.

(i) 'Make plans before you act!' This is a maxim for acting, which appears plausible. But it is, with this degree of universal validity, wrong.

(ii) Planning means that one forms trial sequences of operators and continues to do so until a sequence that (presumably) leads to a goal is found. Now there are many different ways of planning.

(iii) One can, for instance, plan broadly and attempt to include all possible events and possible results of actions taken in the calculations.

(iv) Or else, one can do in-depth planning and establish one line of action from the very beginning. This is risky, as one necessarily has to disregard the side effects and ramifications of the development.

(v) Or one can completely dispense with planning and put one's trust in finding the right measure in a given situation.

Which of these different forms of action preparation is right in any given situation depends on the circumstances. If there is a lot of time pressure and if otherwise everything would be lost, it is reasonable to do some risky planning and 'stake everything on one card'. The probability of losing out is great, but there is simply no time left for behaving differently. If one did nothing at all, everything would be lost anyway, so it is reasonable to prepare at least one action alternative. In a situation in which time pressure is not predominant, it is a better idea to do 'broad' planning and to take as much as possible into consideration. If, however, the environment in which one has to act is chaotic and its laws unknown, one should not plan at all, as it would be a waste of time. It will all happen differently than planned, anyway.

Even if all the operations in the given environment are reversible and any mistakes made are easily undone, it makes sense to do little planning and to act instead. Whether one should plan and how one should plan depends first, on the reversibility of the operators in the given domain of reality, and secondly, on the predictability of the events and ultimately on time pressure. Thus, with regard to planning, the strategic flexibility of thinking consists of one taking into account the features of the situation of action just mentioned and doing broad or in-depth planning, or no planning whatsoever.

The executives are able to judge when which mode of behaviour is appropriate and when it is not. The unsuccessful subjects do not display a sufficient 'fit' between what kind of thinking they employ and the situation. The successful subjects, on the other hand, adapt their thinking to the situation. The good subjects display a greater degree of strategic flexibility than the bad ones.

The reasons we have cited in this paper for the failure of actions in complex and dynamic domains are extremely simple. At the same time, they are very general, and probably one always has to reckon with their effects when people are dealing with complex and dynamic systems. Is it possible to learn how to cope with uncertain, complex, and dynamic network systems? Is it possible to learn how, in general, to cope with uncertainty and complexity? Of course it is possible to learn how one has to act when dealing with the specific demands of the Moros. But the point is whether it is possible to adapt oneself, one's behaviour and thinking, and planning to the possibly contradictory demands of ever changing situations. It is therefore not a question of concrete modes of behaviour, but of whether it is possible to learn to take into account the conditions that show when a concrete mode of behaviour, that is, when this or that form of planning, data gathering, prognosticating, etc. is correct. The fact that this can be learned can be readily seen in the differences between executives and students reported earlier. In terms of pure facts on the Moros, the executives do not know any more than the students. But they know how to adapt themselves to this kind of situation. And they appear to have learned this in the course of their professional lives.

It is possible to learn strategic flexibility. I believe, however, that it is difficult to teach it. It is not a matter of learning a few readily grasped general principles, but of learning a lot of small, 'local' rules, each of which is applicable in a limited area. The point is not to learn how to drive a steamroller with which one can flatten all problems in the same way, but to learn the adroitness of a puppeteer, who at one time holds many strings in his hands and who is able to adapt his movements to the given circumstances in the most sophisticated ways.

I thank Kristin Härtl for many valuable remarks and also Andy Libby for help with the translation of the text.

References

Brehmer, B. & Allard, R. 1986 *Dynamic decision making: a general paradigm and some experimental results*. Uppsala, Sweden: Department of Psychology.

Reichert, U. & Dörner, D. 1988 Heurismen beim Umgang mit einem 'einfachen' dynamischen System. *Sprache Kog.* **7**, 12–24.
Reither, F. 1985 Wertorientierung in komplexen Entscheidungssituationen. *Sprache Kog.* **4**, 21–27.
Schaub, H. & Strohschneider, S. 1989 Die rolle heuristischen wissens beim umgang mit einem komplexen system – oder – sind manager bessere manager? *Forsch. Lehr. Psychol. II.* Memorandum 68.

Discussion

P. Nixon (*Charing Cross Hospital, London, U.K.*). Medical education appears to be designed to reduce the ability of doctors to deal with uncertainty in an environment that is turbulent, uncertain and changing at an unprecedented rate. Their withdrawal, and their inability to provide for the needs of people in such a milieu might be one of the causes of the flight to alternative and complementary medicine.

D. Dörner. I am not familiar with medical education in the U.K., but the reply I am about to give seems to be true of German medical education, and indeed education in general. Training and education tends to provide only standard measures and operations; it does not prepare people for coping with genuine uncertainty or with any kind of situation where standard methods are inappropriate. Such education is always simpler to provide. Indeed, it seems to me very difficult to educate people, by formal means, how to cope with the new and the unknown. On the other hand, our experiments seem to show that people can learn to cope in such situations through successive computer simulations, and perhaps this suggests the possibility of a new kind of training based on computer simulation of tasks with many indefinite and unpredictable variations.

S. D. Rosen (*Charing Cross Hospital, London, U.K.*). Selection, in two respects, is a process that could be applied to Dr Dörner's fascinating study of the relative competence of business executives and students at directing the Moro economy and the executives prowess. My first question is, since we can assume that the business executives are good at business (at least to the extent of being known professionally as business executives), might it not be the case that business executives are selected for their possession of the sorts of skills and attributes that would be effective for running an economy? These skills might well include flexibility, projective planning, etc. A second point (to me one of greater interest) is how far could one use the testing process of the Moro model for the selection of persons with fitting aptitude to professions such as clinical medicine or professional politics? The differentiation of people with these qualities from those solely competent at acquisition of book knowledge could prove invaluable.

D. Dörner. I cannot answer the first question as I don't know how the executives in our experiment were selected, but one should stress the fact, that this Moro problem was not an economic one and therefore skills and attributes, which are effective only for running an economy, would not be very helpful. The executives must possess some rather general skills (which they might have learned when coping with economical problems). The second point is of great interest and importance to me. It might be, that computer simulations of that Moro type might prove to be very effective for the selection of people for positions where it is necessary to cope with uncertain and complex situations.

Phil. Trans. R. Soc. Lond. B. **327**, 475–484 (1990)
Printed in Great Britain

The contribution of latent human failures to the breakdown of complex systems

By J. Reason

Department of Psychology, University of Manchester, Manchester M13 9PL, U.K.

Several recent accidents in complex high-risk technologies had their primary origins in a variety of delayed-action human failures committed long before an emergency state could be recognized. These disasters were due to the adverse conjunction of a large number of causal factors, each one necessary but singly insufficient to achieve the catastrophic outcome. Although the errors and violations of those at the immediate human–system interface often feature large in the post-accident investigations, it is evident that these 'front-line' operators are rarely the principal instigators of system breakdown. Their part is often to provide just those local triggering conditions necessary to manifest systemic weaknesses created by fallible decisions made earlier in the organizational and managerial spheres.

The challenge facing the human reliability community is to find ways of identifying and neutralizing these latent failures before they combine with local triggering events to breach the system's defences. New methods of risk assessment and risk management are needed if we are to achieve any significant improvements in the safety of complex, well-defended, socio-technical systems. This paper distinguishes between active and latent human failures and proposes a general framework for understanding the dynamics of accident causation. It also suggests ways in which current methods of protection may be enhanced, and concludes by discussing the unusual structural features of 'high-reliability' organizations.

1. Introduction

The past few years have seen a succession of major disasters afflicting a wide range of complex technologies: nuclear power plants, chemical installations, spacecraft, 'roll-on-roll-off' ferries, commercial and military aircraft, off-shore oil platforms and railway networks. If we were to focus only upon the surface details, each of these accidents could be regarded as a singular event, unique in its aetiology and consequences. At a more general level, however, these catastrophes are seen to share a number of important features.

(i) They occurred within complex socio-technical systems, most of which possessed elaborate safety devices. That is, these systems required the precise coordination of a large number of human and mechanical elements, and were defended against the uncontrolled release of mass and energy by the deliberate redundancy and diversity of equipment, by automatic shut-down mechanisms and by physical barriers.

(ii) These accidents arose from the adverse conjunction of several diverse causal sequences, each necessary but none sufficient to breach the system's defences by itself. Moreover, a large number of the root causes were present within the system long before the accident sequence was apparent.

(iii) Human rather than technical failures played the dominant roles in all of these accidents. Even when they involved faulty components, it was subsequently judged that appropriate human action could have avoided or mitigated the tragic outcome.

Thanks to the abundance and sophistication of engineered safety measures, many high-risk technologies are now largely proof against single failures, either of humans or components. This represents an enormous engineering achievement. But it carries a penalty. The existence of elaborate 'defences in depth' renders the system opaque to those who control it. The availability of cheap computing power (which provided many of these defences) means that, in several modern technologies, human operators are increasingly remote from the processes that they nominally govern. For much of the time, their task entails little more than monitoring the system to ensure that it functions within acceptable limits.

A point has been reached in the development of technology where the greatest dangers stem not so much from the breakdown of a major component or from isolated operator errors, as from the insidious accumulation of delayed-action human failures occurring primarily within the organizational and managerial sectors. These residual problems do not belong exclusively to either the machine or the human domains. They emerge from a complex and as yet little understood interaction between the technical and social aspects of the system.

Such problems can no longer be solved by the application of still more 'engineering fixes' nor are they amenable to the conventional remedies of human factors specialists. Further improvements in reliability will require more effective methods of risk management. These, in turn, depend upon acquiring a better understanding of the breakdown of complex socio-technical systems, and the development of new techniques of risk assessment. This paper sketches out some of the issues that must be confronted if this ambitious programme is to succeed.

2. Active and latent human failures

Close examination of several recent disasters (especially Bhopal, Challenger, Chernobyl, Zeebrugge and King's Cross) shows the need to distinguish two ways in which human beings contribute to the breakdown of complex systems (see also Rasmussen & Pedersen (1984)).

(i) Active failures: those errors and violations having an immediate adverse effect. These are generally associated with the activities of 'front-line' operators: control room personnel, ships' crews, train drivers, signalmen, pilots, air traffic controllers, etc.

(ii) Latent failures: these are decisions or actions, the damaging consequences of which may lie dormant for a long time, only becoming evident when they combine with local triggering factors (that is, active failures, technical faults, atypical system conditions, etc.) to breach the system's defences. Their defining feature is that they were present within the system well before the onset of a recognizable accident sequence. They are most likely to be spawned by those whose activities are removed in both time and space from the direct human–machine interface: designers, high-level decision makers, regulators, managers and maintenance staff.

Two recent accident investigations, in particular, have dramatically reversed the usual practice of focusing upon the actions of the 'front-line' operators (Sheen, 1987; Fennell 1988). Both the Zeebrugge and King's Cross inquiries concluded that rather than being the main instigators of these disasters, those at the human–machine interface were the inheritors of system defects created by poor design, conflicting goals, defective organization and bad management decisions. Their part, in effect, was simply that of creating the conditions under which these latent failures could reveal themselves.

There is a growing awareness within the human reliability community that attempts to discover and remedy these latent failures will achieve greater safety benefits than will localized efforts to minimize active failures. So far, much of the work of human factors specialists has focused upon improving the immediate human–system interface. Whereas this is undeniably an important enterprise, it only addresses a relatively small part of the total safety problem, being aimed at reducing the active failure tip of the causal iceberg. The remainder of this paper will focus upon latent rather than active failures, beginning with some quantitative evidence from the nuclear power industry.

3. Some data in support of the latent failure argument

The Institute of Nuclear Power Operations (INPO) manages the Significant Event and Information Network for its member utilities both within and outside the United States. In 1985 they issued an analysis of 180 significant event reports received in 1983–84 (INPO 1985). A total of 387 root causes were identified. These were assigned to five main categories: human performance problems, 52%; design deficiences, 33%; manufacturing deficiences, 7%; external causes, 3%; and an 'other unknown' category, 5%.

The human performance problems were further broken down into the following sub-categories: deficient procedures or documentation, 43%; lack of knowledge or training, 18%; failure to follow procedures, 16%; deficient planning or scheduling, 10%; miscommunication, 6%; deficient supervision, 3%; policy problems, 2%; and 'other', 2%.

There are two important conclusions to be drawn from these data. First, at least 92% of all root causes were man-made. Secondly, only a relatively small proportion of the root causes (approximately 8% of the total) were initiated by the operators. The majority had their origins in either maintenance-related activities, or in fallible decisions taken within the organizational and managerial domains.

The major role played by maintenance-related errors in causing nuclear power plant events has also been established by two independent studies (Rasmussen 1980; NUMARC 1985). Of these, simple omissions (the failure to carry out necessary actions) formed the largest single category of identified human problems in nuclear power plant operations.

4. A resident pathogen metaphor

It is suggested that latent failures are analogous to the 'resident pathogens' within the human body, which combine with external factors (stress, toxic agencies, etc.) to bring about disease. Like cancers and cardiovascular disorders, accidents in complex, defended systems do not arise from single causes. They occur through the unforeseen (and often unforeseeable) concatenation of several distinct factors, each one necessary but singly insufficient to cause the catastrophic breakdown. This view leads to a number of general assumptions about accident causation.

(i) The likelihood of an accident is a function of the total number of pathogens (or latent failures) resident within the system. All systems have a certain number. But the more abundant they are, the greater is the probability that a given set of pathogens will meet just those local triggers necessary to complete an accident sequence.

(ii) The more complex, interactive, tightly coupled and opaque the system (Perrow 1984), the greater will be the number of resident pathogens. However, it is likely that simpler systems will require fewer pathogens to bring about an accident as they have fewer defences.

(iii) The higher an individual's position within an organization, the greater is his or her opportunity for generating pathogens.

(iv) It is virtually impossible to foresee all the local triggers, though some could and should be anticipated. Resident pathogens, on the other hand, can be assessed, given adequate access and system knowledge.

(v) It therefore follows that the efforts of safety specialists could be directed more profitably towards the proactive identification and neutralization of latent failures, rather than at the prevention of active failures, as they have largely been in the past.

These assumptions raise some further questions: how can we best gauge the 'morbidity' of high-risk systems? Do systems have general indicators, comparable to a white cell count or a blood pressure reading, from which it is possible to gain some snapshot impression of their overall state of health?

5. A general framework for accident causation

The resident pathogen metaphor is far from being a workable theory. Its terms are still unacceptably vague. Moreover, it shares a number of features with the now largely discredited accident proneness theory, although the pathogen view operates at a systemic rather than at an individual level.

Accident proneness theory floundered when it was established that unequal accident liability was, in reality, a 'club' with a rapidly changing membership (see Reason 1974). In addition, attempts to find a clearly defined accident-prone personality proved largely fruitless.

The pathogen metaphor would suffer a similar fate if it turned out that latent failures could only be identified retrospectively in relation to a specific set of accident circumstances in a particular system. For the analogy to have any value, it is necessary to establish a generic set of indicators relating to system 'morbidity', and then to demonstrate clear connections between these indicators and accident liability across a wide range of complex systems and in a variety of accident conditions.

In what follows, an attempt will be made to develop the pathogen metaphor into a theoretical framework for considering the aetiology of accidents in complex technological systems. The challenge is not just to provide an account of how active and latent failures combine to produce accidents, but also to show where and how more effective remedial measures might be applied.

Before considering the pathology of complex systems, we must first identify their essential, 'healthy' components. These are the basic elements of production. All complex technologies are involved in some form of production, whether it be energy, a chemical substance, or the mass transportation of people by land, sea or air. There are five basic elements to any productive system: decision makers, line management, preconditions, productive activities and defences.

(i) Decision makers. These include both the architects and the senior executives of the system. Once in operation, the latter set the production and safety goals for the system as a whole. They also direct, at a strategic level, the means by which these goals should be met. A large part of their function is concerned with the allocation of finite resources. These comprise money, equipment, people and time.

(ii) Line management. These are the departmental specialists who implement the strategies of the decision makers within their particular spheres of operation: operations, training, sales, maintenance, finance, safety, engineering support, personnel, and so on.

(iii) Preconditions. Effective production requires more than just machines and people. The equipment must be reliable and of the right kind. The workforce must be skilled, alert, knowledgeable and motivated.

(iv) Productive activities. These are the actual performances of machines and people: the temporal and spatial coordination of mechanical and human activities needed to deliver the right product at the right time.

(v) Defences. Where the productive activities involve exposure to hazards, both the human and mechanical components of the system need to be provided with safeguards sufficient to prevent forseeable injury, damage or costly outages.

The human contributions to accidents are summarized in figure 1. They are linked there to each of the basic elements of production, portrayed as 'planes' lying one behind the other in an ordered sequence. The question at issue is: how do fallible decisions translate into unsafe acts capable of breaching the system's defences?

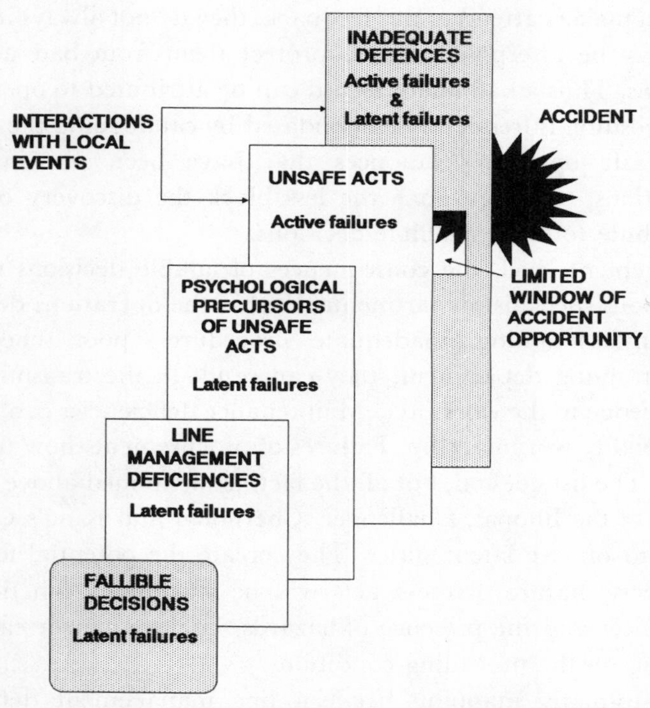

FIGURE 1. Showing the relationship between the various human contributions to accidents and the basic elements of production. Latent failures have their primary systemic origins in the fallible decisions made by senior executives. They are subsequently translated into different forms as the effects of these decisions pass through the system during the production process.

It is assumed that latent failures (resident pathogens) have their primary systemic origin in the errors of high-level decision makers. But they are also introduced into all levels of the system by the human condition. Error proneness and the capacities for being stressed, failing to perceive hazards, being ignorant of the system, and having less than ideal motivation are brought by each individual into the workplace.

Even in the best run organizations, a significant number of influential decisions will subsequently prove to be mistaken. Fallible decisions are an inevitable part of the design and management process. The issue is not so much how to prevent them, but how to ensure that their adverse consequences are detected and recovered.

All organizations must allocate limited resources to two distinct goals: production and safety. In the long term, these are clearly compatible. But short-term conflicts of interest will arise in which the resources given to production could diminish safety, and conversely. There are a number of reasons why these dilemmas will tend to be resolved in favour of production rather than safety goals.

First, resources directed at improving productivity have relatively certain outcomes; those aimed at enhancing safety do not, at least in the short term (Brehmer, 1988). This is due to the large part played by stochastic factors in accident causation.

Secondly, the feedback generated by the pursuit of production goals is generally unambiguous, rapid, compelling and (when the news is good) highly reinforcing. In sharp contrast, that derived from the pursuit of safety goals is largely negative, intermittent, often deceptive and perhaps only compelling after a major accident or a string of incidents.

Even when decision makers attend to this feedback, they do not always interpret it correctly. Defensive 'filters' may be interposed, which protect them from bad news and encourage extrapunitive reactions. Thus a bad safety record can be attributed to operator carelessness or incompetence. This position is frequently consolidated by cataloguing the various engineering safety devices and safe operating practices that have been put in place. These are understandable reactions, but they none the less block the discovery of effective counter-measured and contribute to further fallible decisions.

At the line management level, the consequences of fallible decisions manifest themselves differently in the various specialist departments. Within the operations department, they can take the form of undermanning, inadequate procedures, poor scheduling and unsafe assignments. In the training department, they can result in the transmission of inadequate skills, rules and knowledge to the workforce. Maintenance deficiencies can reveal themselves in poor planning and shoddy workmanship. Failures of procurement show up as dangerous and defective equipment. The list goes on, but all the factors mentioned above played a significant part in the aetiology of the Bhopal, Challenger, Chernobyl and King's Cross disasters.

Psychological precursors are latent states. They create the potential for a wide variety of unsafe acts. The precise nature of these acts will be a complex function of the task, the environmental influences and the presence of hazards. Each precursor can give rise to many unsafe acts, depending on the prevailing conditions.

There is a many-to-many mapping between line management deficiencies and these psychological precursors. Failures in the training department, for example, can translate into a variety of precursors: high workload, undue time pressure, inappropriate preception of hazards, ignorance of the system and motivational difficulties. Likewise, any one precondition (for example, undue time pressure) could be the product of many line management deficiencies: poor sheduling, inadequate procedures, inappropriate training and maintenance failures.

A useful way of thinking about these transformations is as types converting into tokens. Deficient training is a general failure type that can reveal itself, at the precursor level, as a variety of pathogenic tokens. Such a view has important remedial implications. Rectifying a failure type could, in principle, eliminate a large class of tokens. The type–token distinction is a hierarchical one. Precondition tokens at the precursor level become types for the creation of tokens at unsafe act level.

A psychological precursor, either alone or in combination with others, can play a major role in provoking and shaping an almost infinitely large set of unsafe acts. The stochastic character of this onward mapping reveals the futility of 'tokenism': the concentration of remedial efforts upon preventing the recurrence of specific unsafe acts. Although certain of these acts may fall into a recognizable category (for example, failing to wear personal safety equipment) and so be amenable to targeted safety programmes, the vast majority of them are unforeseeable and occasionally quite bizarre.

This view of accident causation suggests that unsafe acts are best reduced by eliminating their psychological precursors rather than the acts themselves. However, it must be accepted that whatever measures are taken, some unsafe acts will still occur. It is therefore necessary to provide a variety of defences to intervene between the act and its adverse consequences. Such defences can be both physical and psychological. The latter are as yet relatively unexploited, and involve procedures designed to improve error detection and recovery.

Very few unsafe acts will result in damage or injury. In a highly protected system, the probability that the consequences of an isolated action will penetrate the various layers of defence is vanishingly small. Several causal factors are required to create a 'trajectory of opportunity' through these multiple defences. Many of the causal contributions will come from latent failures in the organizational structure, or in the defences themselves. Others will be local triggering factors. These could be a set of unsafe acts committed during some atypical (but not necessarily abnormal) system state. Examples of the latter are the unusually low temperature on the night preceding the Challenger launch, the voltage–generator tests carried out just before the annual maintenance shut-down in the Chernobyl-4 reactor, and the nose-down trim of the Herald of Free Enterprise because of a combination of unusually high tide and unsuitable docking facilities.

A significant number of accidents in complex systems arise from the deliberate or unwitting disabling of defences by operators in pursuit of what, at the time, seem to be sensible or necessary goals. The test plan at Chernobyl required that the emergency core cooling system should be switched off, and the need to improvise in an unfamiliar and increasingly unstable power regime later led the operators to strip the reactor of its remaining defences. At Zeebrugge, the overworked and undermanned crew of the Herald of Free Enterprise left harbour with the bow doors open. This was an oversight caused by a bizarre combination of active failures (Sheen 1987), but it was also compounded by strong management pressures to meet the stringent schedule for the Dover docking.

6. Managing safer operations

An effective safety information system has been found to rank second only to top management concern with safety in discriminating between safe and unsafe companies, matched on other variables (Kjellen 1983). The feedback loops and indicators that could go to make up such a system are shown in figure 2.

Loop 1 (reporting accidents, lost time injuries, etc.) represents the minimum requirement.

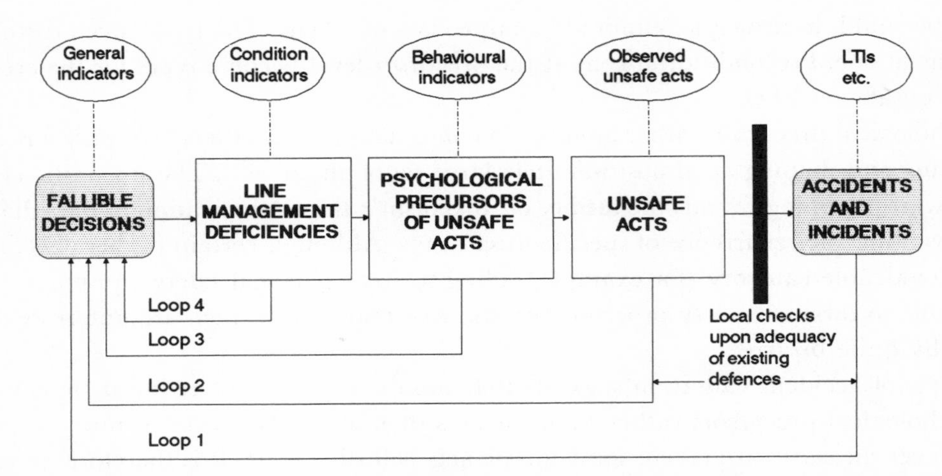

FIGURE 2. The actual and potential feedback loops and indicators associated with each of the basic elements of production. Loop 1 already exists in most systems, and communicates information about accidents, lost time injuries, etc. Loop 2 is potentially available through Unsafe Act Auditing. Loops 3 and 4 could provide information regarding latent failures; though, in practice they are rarely seen in place.

In most cases, however, the information supplied is too little and too late for effective proactive control. The events that safety management seeks to prevent have already occurred.

Loop 2 is potentially available through unsafe act auditing procedures. In practice, however, this information is usually only disseminated to the lower, supervisory levels of the organization.

The main thrust of the present view of accident causation is towards the establishment of loops 3 and 4. It argues that the most effective way of managing safety is by acting upon types rather than tokens; that is, by influencing system and individual states occurring early on in the history of a possible accident. To identify these indicators, and to find ways of neutralizing the general failure types so revealed constitute the major challenges facing contemporary accident researchers. For the moment, however, we will consider only the most global of these diagnostic signs: the general indicators associated with top-level decision making.

The general indicators shown in figure 2 cover two broad aspects of a system's safety management. The first relates to the variety and sensitivity of its feedback loops. The second deals with the senior executives' responses to safety-related data. No amount of feedback will enhance system safety if the information supplied is not acted upon in a timely and effective manner.

Westrum (1988) has provided a useful classification of the ways in which organizations differ in their responses to safety-related information. These reactions fall into three groups: denial, repair and reform actions.

(i) Denial actions. These may take one or both of the following forms: suppression, 'whistleblowers' are punished or dismissed and their observations removed from the record; and encapsulation, the observers are retained, but the validity of their observation is disputed or denied.

(ii) Repair actions. Externally, these can take the form of a public relations exercise in which the observations are allowed to emerge, but in a reassuring and sugar-coated manner. Internally, the problem is admitted, but it is only addressed at a local level. 'Offending' operators are disciplined or relocated. Dangerous items of equipment are modified to prevent the recurrence of a specific kind of observed failure. The wider implications of the problem are denied.

(iii) Reform actions. These take two forms: dissemination, the problem is admitted to be global, and global action is taken upon it; reorganization, action on the problem leads to a fundamental reappraisal and reform of the system as a whole.

The more effective the organization, the more likely it is to respond to safety data with reform actions. Less adequate organizations will confine themselves to either denial or repair actions. This leads to a tripartite classification of organizations.

(i) Pathological organizations possess inadequate safety measures, even under normal circumstances. They habitually sacrifice safety for greater productivity, often under severe economic pressure, and they actively circumvent safety regulations.

(ii) Calculative organizations try to do the best job they can by using 'by-the-book' methods. These are usually adequate under normal operating circumstances, but often fail to thwart the development of the multiple-cause accidents, discussed earlier.

(iii) Generative organizations set safety targets for themselves beyond ordinary expectations, and fulfill them because they are willing to do unusual things in unconventional ways. They emphasise results more than methods, and value substance more than form. Notable exemplars of this last category have been investigated by La Porte and his colleagues at the University of California, Berkeley.

7. High-reliability organizations

The La Porte group (see La Porte & Consolini 1989) has made an intensive study of three high-reliability organizations: the Federal Aviation Authority's air traffic control system, Pacific Gas and Electric's power generating system and two U.S. Navy nuclear aircraft carriers. These organizations share at least two goals: to avoid altogether major failures that could cripple or even destroy, the system; and to cope safely with periods of very high peak demand and production whenever they arise. All of these organizations perform complex and hazardous tasks under considerable time pressure, and they do so with a very low error rate and an almost total absence of catastrophic failure. What are the ingredients? Can we 'bottle' them?

Perhaps the most significant feature of these organizations is their complex yet highly adaptive structural reactions to changing levels of hazard. Each organization has three distinct authority structures: routine, high-tempo and emergency. Each structure has its own characteristic practices, communication pathways and leadership patterns.

The routine mode reveals the familiar hierarchical pattern of rank-dependent authority. This is the face of the organization most evident to the casual observer. It functions with the use of extensive 'standard operating procedures'.

Just beneath the surface of this bureaucratic structure is the high-tempo mode, practised by the same individuals, but in quite a different manner. Authority is no longer based upon rank, but upon functional skills. Formal status defers to expertise. Communications switch from largely vertical channels to being richly horizontal among task-related groups.

Within these high-tempo groups, the La Porte team noted an extraordinary sensitivity to the incipient overloading of any one of its members. For example, when an air traffic controller has an unusually large number of aircraft on his screen, supervisors and other controllers will gather around silently and watch for danger points. When found, they are shown by pointing at the screen. Few words are spoken. When the load has eased, the impromptu support group fades away as quietly as it arrived.

The emergency mode is triggered by unequivocal signs that a well-defined danger is imminent. Authority patterns in this mode are based upon a preprogrammed and well-rehearsed allocation of duties. Individuals regroup themselves into different functional units on the basis of a predetermined plan, tailored to the particular nature of the emergency. La Porte and Consolini (1989) comment upon these co-existing structures as follows: 'Contemporary organization theory literature does little to alert one to the likelihood of these multi-layered, nested authority systems. We are familiar with different types of organization that parallel each of these modes. There are bureaucratic organizations, professional ones, and disaster response ones. We have not thought that all three might be usable by the same organizational membership.'

Can we build these adaptive structural ingredients into high-risk organizations at their outset, or must they evolve painfully and serendipitously over many years of hazardous operating experience? It is probably too early to tell. But it is clear that a close study of high-reliability organizations should feature prominently on the research agendas of those concerned with understanding and preventing the kinds of disaster discusssed in this paper. Just as in medicine, it is probably easier to characterize sick systems rather than healthy ones. Yet we need to pursue both of these goals concurrently if we are to understand and then create the organizational bases of system reliability.

Many people contributed to the ideas expressed in this paper. Two, in particular, deserve special mention. Patrick Hudson of the University of Leiden was the first to apply the type–token distinction in the context of accident causation. The production-based framework for accident causation emerged from discussions with John Wreathall of Science Applications International Corporation (Columbus, Ohio).

References

Brehmer, B. 1988 Changing decisions about safety in organizations. World Bank Workshop on Safety Control and Risk Management, 18–20 October, Washington, D.C.

Fennell, D. 1988 *Investigation into the King's Cross underground fire*. Department of Transport, London: HMSO.

INPO 1985 *An analysis of root causes in 1983 and 1984 significant event reports*. Atlanta, Georgia: Institute of Nuclear Power Operations.

Kjellen, U. 193 *The Deviation Concept in Occupational Accident Control, TRITA/AOG-0019, Arbetsolycksfalls Gruppen*. Stockholm: Royal Institute of Technology.

La Porte, T. R. & Consolini, P. M. 1988 Working in practice but not in theory: theoretical challenges of high reliability organizations. Annual Meeting of the American Political Science Association, 1–4 September, Washington, D.C.

Numarc 1985 *A maintenance analysis of safety significant events*. Nuclear Utility Management and Human Resources Committee. Atlanta, GA: Institute of Nuclear Power Operations.

Perrow, C. 1984 *Normal accidents: living with high risk technologies*. New York: Basic Books.

Rasmussen, J. & Pedersen, O. M. 1983 Human factors in probabilistic risk analysis and risk management. In *Operational safety of nuclear power plants*, vol. 1. Vienna: International Atomic Energy Agency.

Rasmussen, J. 1980 What can be learned from human error reports. In *Changes in working life* (ed. K. Duncan, M. Gruneberg & D. Wallis). London: Wiley.

Reason, J. T. 1974 *Man in motion*. London: Weidenfeld.

Sheen, Mr Justice. 1987 *Herald of Free Enterprise*. Report of Court no. 8074. Department of Transport. London: HMSO.

Westrum, R. Organizational and inter-organizational thought. World Bank Workshop on Safety Control and Risk Management, 18–20 October, Washington, D.C.

Phil. Trans. R. Soc. Lond. B **327**, 485–492 (1990)

Printed in Great Britain

Auditory warning sounds in the work environment

By R. D. Patterson
MRC Applied Psychology Unit, 15 Chaucer Road, Cambridge CB2 2EF, U.K.

One of the most common auditory warnings is the ambulance 'siren'. It cuts through traffic noise and commands one's attention, but it does so by sheer brute force. This 'better safe than sorry' approach to auditory warnings occurs in most environments where sounds are used to signal danger or potential danger. Flooding the environment with sound is certain to attract attention; however it also causes startled reactions and prevents communications at a crucial point in time. In collaboration with several companies and government departments, the MRC Applied Psychology Unit performed a series of auditory warning studies. The main conclusions of the research were that the number of immediate-action warning sounds should not exceed about six, and that each sound should have a distinct melody and temporal pattern. The experiments also showed that it is possible to predict the optimum sound level for a warning sound in most noise environments.

Subsequently, a set of guidelines for the production of ergonomic auditory warnings was developed. The guidelines have been used to analyse the environments in both fixed-wing and rotary-wing aircraft, and to design prototype warning systems for environments as diverse as helicopters, operating theatres and the railways.

1. Introduction

For some time now it has been clear both to aircrew and the Civil Aviation Authority (CAA) that the auditory warning systems on civil aircraft are, to say the least, non-optimal. Briefly, the pilots complained that there were too many warning sounds, that they were too loud, and that they were confusing. The problems are illustrated in an incident report filed with CHIRP, a confidential incident reporting service provided by the Institute of Aviation Medicine at Farnborough. The following is an abbreviated version of the first few paragraphs of the report:

> 'I was flying in a Jetstream at night when my peaceful reverie was shattered by the stall audio warning, the stick shaker, and several warning lights. The effect was exactly what was NOT intended; I was frightened numb for several seconds and drawn off instruments trying to work out how to cancel the audio/visual assault rather than taking what should be instinctive actions.
>
> The combined assault is so loud and bright that it is impossible to talk to the other crew member, and action is invariably taken to cancel the cacophony before getting on with the actual problem.'

A review of the existing situation immediately revealed that all three of the pilots' complaints were justified. Some of the aircraft had as many as 15 auditory warnings; they were not conceived as a set and there was no internal structure to assist the learning and retention of the warnings. A number of the warnings produced sound levels over 100 dB at the pilot's ear and virtually all of the warnings came on instantaneously at their full intensity. With regard to confusion, there appeared to be no order among the different spectra and the different temporal patterns used in the warning sounds. Furthermore, it appeared that if two of the warnings came on simultaneously, they would produce a combined sound that would make it difficult to identify either of the conditions involved.

The pilots' problems occur over and over again in noisy, high-workload environments where auditory warnings are used to signal danger. An extreme example of the confusion problem occurs in the intensive care wards of large hospitals. There may be as many as six critically injured patients in an intensive care ward, and each of the patients is surrounded by life-support equipment and monitors that may present 10 or more auditory warning sounds. Thus there can be more than 60 relevant auditory warnings in one ward, far more than the staff could hope to learn and remember. What is more, for reasons of economy, most of the warning sounds are high-frequency tones that differ only in their frequency and intensity. The auditory system of humans is not designed to preserve the absolute frequency or intensity of sound sources, rather it is designed to listen for changes in sounds. Furthermore, high-frequency tones are not localizable. Thus it was entirely predictable on theoretical grounds alone, that there would be significant confusions between the warning sounds in this environment (Patterson *et al.* 1986). With regard to level, excessively loud warnings are not restricted to aircraft. Warning horns and fire bells that were appropriate on steam locomotives with open cabs, are still present in the cabs of modern electric trains. Although the history of existing auditory warning systems is fascinating, the purpose of this paper is to explain how we can produce better auditory warning systems.

Before proceeding, it is perhaps worth pointing out why auditory warnings are useful and why they are necessary. Warning sounds are useful because hearing is a primary warning sense. It does not matter whether operators are concentrating on an important visual task, or relaxing with their eyes closed; either way, if a warning sound occurs it will be detected automatically and routed through on a priority line to the brain. The need for auditory warnings, in addition to visual warnings, is exemplified in an accident report which describes how a passenger helicopter descended inadvertently into the sea (Cooper 1984). Until this accident helicopters did not have auditory warning sounds. In this particular case it appears that the pilots were busy looking for landfall in fog and they simply did not see the flashing yellow light behind the control column telling them that they were descending to a dangerously low height. An auditory warning sound, as the report concludes, would almost undoubtedly have averted this accident. Thus the question is not one of whether we should, or should not, have warning sounds. Rather the question is whether we can construct sets of warning sounds that get one's attention reliably without causing startled reactions.

At the Applied Psychology Unit we began by concentrating on four problems: (*a*) what is the correct level for a warning sound, that is, the level that will render the warning reliably audible but not adversely loud? (*b*) what are the appropriate spectral characteristics to ensure that a warning is not only audible but also discriminable from other members of the set? (*c*) what are the short-term temporal characteristics that make a warning sound arresting without producing a startle response? (*d*) what are the longer-term temporal characteristics that give the warning sound a distinctive and memorable rhythm? The answers to these questions, and the guidelines that summarise the answers, form the topic of the remainder of this paper.

2. The appropriate sound level for auditory warnings

As a guideline, if a warning is to be clearly audible and draw the attention of the flight-crew reliably, the spectral components of the warning sound must be 15 dB above the threshold imposed by the background noise on the individual spectral components. Thus the problem of determining the appropriate range for the components of a warning sound reduces to one of finding the threshold imposed by the noise background on the components. The spectrum for background noise on the flight deck of a Boeing 727 flying at 240 knots is shown by the bottom curve in figure 1.

Threshold for any particular component is largely determined by the noise components in the same region of the spectrum as the warning component itself. It is as if the observer were listening for the warning component through an auditory filter centred at the frequency of the component; that is, threshold is determined by the amount of noise power that leaks through that filter with the warning component. After Patterson (1982), threshold was calculated as a function of frequency, at multiples of 0.01 kHz, and it is shown by the smooth curve in the centre of figure 1. As the noise spectrum is smooth, the threshold function is also smooth and follows the noise spectrum fairly closely. Note, however, that threshold does actually diverge from the noise spectrum as frequency increases because the bandwidth of the filter increases with its centre frequency.

The individual spectral components of an auditory warning must be above the threshold curve to be just audible in that noise. To be reliably audible four or more components of a warning should be at least 15 dB above auditory threshold. The lower boundary of the banded area above the threshold curve is 15 dB above threshold and so it represents the lower bound of the appropriate range for auditory-warning components. In the frequency range 1.0–2.0 kHz, the curve shows that the background noise requires the use of warning components in excess of 85 dB SPL. As this is already a high level it is recommended that the upper bound of the appropriate range for warning components should be limited to about 10 dB above the lower bound. Thus the banded region is the portion of the space in which the auditory warning components should fall.

The spectral components of the configuration horn of the Boeing 727 are shown by the dashed vertical lines in figure 1. The largest component is more than 20 dB above the

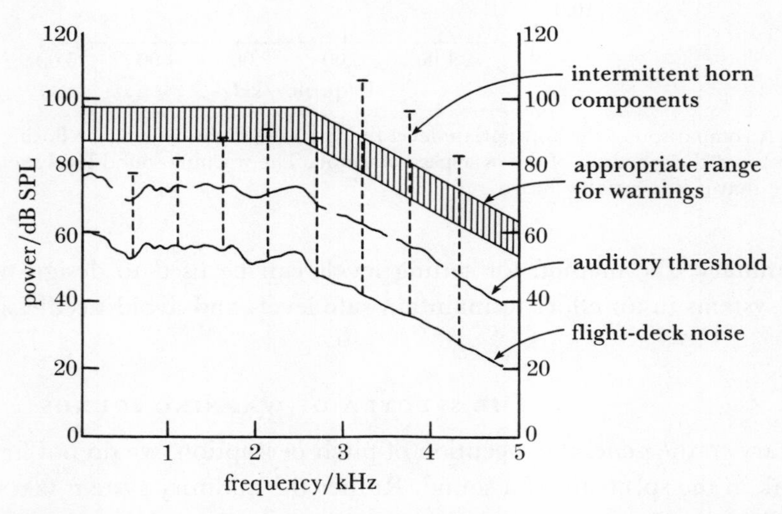

FIGURE 1. The range of appropriate levels for warning sound components on the flight-deck of the Boeing 727 (vertical-line shading). The minimum of the appropriate-level range is approximately 15 dB above auditory threshold (broken line) which is calculated from the spectrum of the flight-deck noise (solid line). The vertical dashed lines show the components of the intermittent warning horn, some of which are well above the maximum of the appropriate-level range.

maximum of the appropriate level range. Sounds in excess of 85 dB SPL are aversive for most people, particularly in this frequency region. In a separate study (Patterson *et al.* 1985) a configuration horn was presented to two test pilots in flight at six different sound levels. They independently chose the version of the warning horn that set the main components in the appropriate-level range. There is, then, good agreement between the level that the model of auditory masking shows is required for reliable detection, and the level that the test pilots felt was appropriate.

This method of specifying the level for auditory warnings has also been used to review the loudness of the warnings on a Boeing 747 (Patterson *et al.* 1985). Eight of the eleven warning sounds had acceptable sound levels, with five of the warnings having near optimum levels. However three of the warnings were actually found to be too soft: the overspeed, landing-configuration, and take-off configuration warnings. The main spectral peak of the overspeed warning is just under the lower bound of the appropriate level range (figure 2). As the warning must be available in the loudest of the noise environments on this aircraft, the level of the warning should be increased by 8–10 dB so that it is nearer the maximum of the appropriate-level range.

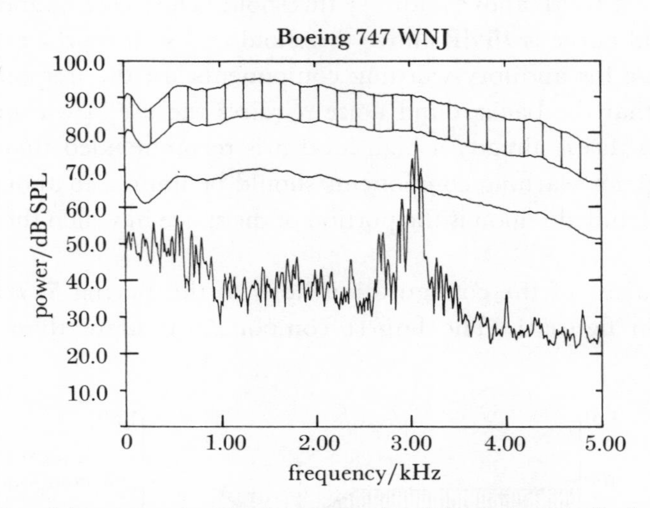

FIGURE 2. A comparison of the appropriate-level range on the flight-deck of the Boeing 747 (the vertically shaded region) and the spectrum of the overspeed warning. The warning sound level should be raised about 8 dB to ensure that it is invariably heard.

In summary, the method for setting levels can be used to design new systems or modify existing systems in an effort to maintain safe levels and avoid needlessly annoying levels.

3. The spectra of warning sounds

Contrary to the general conception of pitch perception, we do not hear a separate pitch for each peak in the spectrum of a sound. Rather the auditory system takes the information from temporally related components and maps them back onto one perception, namely, a pitch corresponding to the fundamental of the harmonic series implied by the related components. This property of the hearing mechanism has important implications for the design of auditory warnings. Specifically it enables us to design warnings that are highly resistant to masking by spurious noise sources; that is, resistant to masking by those unpredictable noise sources that might accompany the event that causes an emergency on an aircraft. Briefly, a warning sound that has four or more components in the appropriate level range, and which are spread across the spectrum, is much much less likely to be masked by a spurious noise source than one in which all of the energy is concentrated at one harmonic.

On the Boeing 747 the warning sounds chosen to represent glide slope, passenger evacuation, and overspeed problems have only one spectral peak at about 0.35, 3.6 and 3.1 kHz, respectively (Patterson *et al.* 1985). The overspeed spectrum is shown in figure 2. Some of the turbines and pumps on aircraft could produce noise in the 3–4 kHz region should they become worn, and it could make the warnings difficult to hear. In contrast, the harmonic content of the intermittent horn in figure 1 is excellent. It is very unlikely that a spurious masker could prevent its being heard, even after the level has been adjusted.

4. The temporal characteristics of warning sounds

A study was performed at APU to determine whether flight-deck warnings are confusing, that is, whether they are intrinsically difficult to learn and remember (Patterson 1982). Groups of naive listeners were taught to recognize a set of ten auditory warnings drawn from the flight-decks of a variety of current civil aircraft. The results show that the first four to six warnings are acquired quickly, but thereafter, the rate of acquisition slows markedly. There is no inherent difficulty in learning warning sounds, but beyond the first six it does require appreciatively more effort. The results of the learning and retention study cannot be directly applied to pilots on the flight-deck. They do, however, reinforce the growing belief that aircraft with sets of ten or more warnings have too many. On the other hand, the ease with which naive listeners learn six arbitrary warnings suggests that a set of six should prove entirely reliable.

A confusion analysis applied to the errors made during the learning phase of the experiment showed that warnings with the same pulse-repetition-rate were likely to be confused even when there were gross spectral differences between the warning sounds. It is important to stress that the listeners were naive and that their rate of confusion is very high with respect to the rate that might be expected to occur on the flight-deck. The results did, however, show that any potential for confusion could be dramatically reduced by employing a richer variety of temporal patterns, that is, distinctive rhythms. Subsequently, new sets of warnings for helicopters (Lower *et al.* 1986; Patterson *et al.* 1989) and have included rhythm distinctions, and learning tests on these new warning sets reveal much lower confusion rates.

Changes in sound level are useful for drawing a listener's attention, and the greater the rate of change, the more demanding the sound. However, some of the existing flight-deck warnings go from off to full on at a level over 100 dB SPL in under 10 ms. In the natural environment a rapid rise to a high sound level is characteristic of a catastrophic event in the listener's immediate surroundings. The natural response to such an event is an involuntary startle reflex in which the muscles are tensed in preparation for a blow or a quick response. Instantaneous responses often prove incorrect, and so they are specifically discouraged on the flight-deck and in pilot training. The abrupt onsets of current warnings are not justified by a requirement for fast motor responses. Patterson (1982) has suggested a rise time of 20 ms. The duration of the offset of the pulse is determined by the same factors.

5. A prototype ergonomic warning sound

The temporal structure of a warning that might be used in a train, plane or hospital is shown in figure 3. The upper row shows the basic pulse with its rounded onsets and offsets. The middle row shows the burst, or pulse pattern, used to represent the warning. The bottom row shows the timecourse of the complete warning. The waveform within the pulse is unique to a particular warning; it carries the spectral information of the warning sound and is never altered. In this case, a burst of six pulses defines the warning sound. The basic grouping of four, clustered pulses followed by two, irregularly spaced pulses provides the rhythm of the sound which, combined with the spectral characteristics stored in the waveform, gives the sound its distinctiveness. The spacing of the pulses is varied within the burst to vary the perceived urgency.

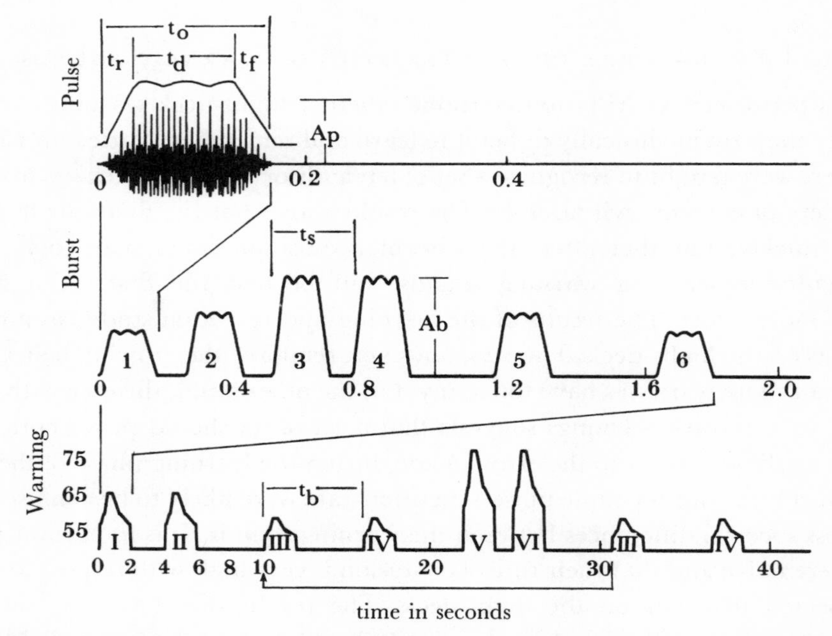

FIGURE 3. The modules of a prototype warning sound: the sound pulse is an acoustic wave with rounded onsets and offsets and a distinctive spectrum; the burst is a set of pulses with a distinctive rhythm and pitch contour; the complete warning sound is a set of bursts with varying intensity and urgency.

An arresting warning can be produced without risking a startle reaction by bringing the warning on at a comparatively low level and increasing the level of successive pulses quickly as shown in the middle row. This amplitude envelope gives the impression that an object is moving forwards rapidly and then receding slowly, and this apparent motion draws attention. At the same time, since the first pulse comes on at a moderate level, the warning does not cause a startle reaction. The basic pulse is similarly given a rounded top rather than an abrupt onset or offset to reduce the risk of a startle reaction.

The timecourse of the complete warning sound is shown in the bottom row of figure 3. The little 'houses' designated by Roman numerals, represent different versions of the burst that vary in their urgency. The height of the houses shows the relative intensity of the bursts. The spectral and temporal characteristics of the pulse (upper row) and the burst (middle row) give the warning sound its distinctive character. The pitch, intensity, and speed of the burst are used to vary the perceived urgency of the warning sound. A burst can be thought of as a brief atonal melody with a syncopated rhythm.

When the situation necessitates, the warning sound comes on and the first burst is played at a pitch and speed that indicate moderate urgency, and at a level that is clearly audible but not excessive, as determined by the background noise in the environment. Then the burst is repeated. At this point, it is highly likely that the warning has conveyed its message, and that further repetition of the burst in its urgent form would be needlessly irritating. At this point, then, the pitch, level and speed of the burst are lowered to reduce its perceived urgency, and it is played every 4 s or so in this non-urgent form (III and IV). With the level reduced and the time between bursts extended, one can communicate verbally in the presence of the warning without difficulty, an important advantage in an emergency. If the condition that initiated the warning sound is not alleviated within a reasonable length of time, the warning returns in its most urgent form, conveyed by a pair of bursts (V and VI) with a relatively high pitch, a fast pulse rate, and a level that overrides any ongoing speech and commands attention. Then, the warning returns to the background level to permit communication. Bursts III–VI are repeated until the condition that initiated the warning is corrected, or until someone indicates their attendance. In the latter case, the warning remains on in the background form with the urgent bursts repressed and the non-urgent bursts repeating every four seconds or so to show that an abnormal condition still exists.

6. Applications

Five sets of auditory warnings have now been developed according to the principles outlined to this point. In the case of the Civil Aviation Authority the purpose of the set was to illustrate the guidelines (Patterson 1982) and the kind of warning sounds that would satisfy a British Aircraft Standard that was being developed in response to pilots' complaints. In the case of hospitals, the purpose of the warning set was to show to British and International Standards organizations the kind of civilised, distinctive warning sounds that could be specified in a Standards document and used to replace the cacophony of buzzers and bells used currently in operating theatres and intensive care wards (Patterson *et al.* 1986). The Standards specify, and the demonstration warnings illustrate, two forms of hospital warning system. In one case there are only three sounds, each of which shows a whole category of problems and which are differentiated by their urgency. In the second form, the three category sounds are supplemented by six specific warnings all of which show urgent conditions occurring in the topmost category. The design represents an ingenious compromise that enables each authority or hospital ward to tailor the system to their own needs by adding a small number of the highly urgent warning sounds to the general set of three category warnings.

Three related sets of warnings were designed for use in helicopters (Lower *et al.* 1986). The first set was developed for the multi-role Sea King helicopter and consisted of ten warning sounds. A learning experiment was performed with helicopter pilots and it showed that the new sounds were much more resistant to confusion than those being used currently in civil airliners. The second set of warnings was produced for the Lynx helicopter and used to check the guidelines for setting the sound levels of the warning sounds. Detection levels were measured in a Lynx helicopter shell and it showed that the model of auditory masking was as accurate as the noise measurements that could be made beside the pilot's ear. The third set of warnings was produced for helicopters ferrying staff and supplies to North Sea oil rigs. A subset of the warnings is currently installed in more than 150 North Sea helicopters.

Finally, at the request of British Rail Research, a set of warning sounds was designed for use by trackside maintenance crews to warn of approaching trains. In this case, the aim was to preserve the correspondence between trackside warning function and the sound that British Rail already had (the PeeWee). All of the new warning sounds were constructed from components of the existing sound (Patterson *et al.* 1989). Furthermore, all four warning sounds had to be audible in the presence of no less than 46 different noise environments.

7. Conclusion

In summary, one can now design and build warning sounds that present their message with appropriate urgency and promptly fall back to permit vital communication, returning to interrupt forcefully only if there is reason to believe that the condition is not receiving sufficient attention.

The author thanks Robert Milroy and Mike Lower for their continued assistance in producing and testing the auditory warnings over the years of this project.

References

Cooper, D. A. 1984 Report on the accident to British Airways Sikorsky S-61N, G-BEON. Department of Transport Aircraft Accident Report 8/84.

Lower, M. C., Patterson, R. D., Rood, G., Edworthy, J., Shailer, M. J., Milroy, R., Chillery, J. & Wheeler, P. D. 1986 The design and production of auditory warnings for helicopters 1: the Sea King. *Inst. Sound Vib. Res. Rep.*, no. AC527A.

Patterson, R. D. 1982 Guidelines for auditory warning systems on civil aircraft. Civil Aviation Authority. Paper 82017.

Patterson, R. D., Milroy, R. & Barton, J. 1985 Auditory warnings on the BAC 1-11 and the Boeing 747. Civil Aviation Authority. Paper 85004.

Patterson, R. D., Cosgrove, P., Milroy, R. & Lower, M. C. 1989 Auditory warnings for the British Rail Inductive Loop Warning System. *Proc. Inst. Acoust.* **II**, 51–58.

Patterson, R. D., Edworthy, J., Shailer, M. J., Lower, M. C. & Wheeler, P. D. 1986 Alarm Sounds for medical equipment in intensive care areas and operating theatres. *Inst. Sound Vib. Res. Rep.*, no. AC598.

Discussion

T. F. Mayfield (*Rolls Royce and Associates, U.K.*). Dr Patterson has presented what seems to be a large number of discrete tone combinations. Is it suggested that all these would be used in any one situation.

R. D. Patterson. For a discussion of the appropriate number of warning sounds, see §4.

Phil. Trans. R. Soc. Lond. B **327**, 493–502 (1990)
Printed in Great Britain

Effective decisions and their verbal justification

BY D. E. BROADBENT, F.R.S.
Department of Experimental Psychology, University of Oxford, South Parks Road, Oxford OX1 3UD, U.K.

Recent experimental work on human control of complex systems has drawn attention to the discrepancy that may exist between the person's reported knowledge of the system and their ability to control it. Sometimes people act correctly but cannot answer questions about what they are doing; sometimes they can say verbally what they should do (perhaps having had verbal instruction in the right answers), but still do not do it. This discrepancy is of major practical importance, for example in designing training programmes or in eliciting expert knowledge for incorporation in a mechanical 'expert system'. It is also puzzling for psychological theory, as it rules out certain plausible models of the functioning of the brain. This paper considers what mechanisms are still possible.

1. INTRODUCTION

The practical problem

Suppose we consider some practical task that involves human beings controlling a complex dynamic process. Why should we be concerned about the links between their performance and their ability to tell us what they are doing? As some might argue, all that matters is that their handling of the system should be safe and efficient. From that point of view, worrying about their conscious experience (whatever that means) is a problem for the philosopher or perhaps the psychiatrist, not something that need occupy the safety engineer or the assessor of human reliability. That however would be a mistake; there are at least three reasons why we should think hard about the connection between words and actions.

First, there is the question of training. A safe and competent operator of a nuclear power station, or a chemical plant, or an air traffic control system, must clearly possess a substantial database of knowledge about physics, or chemistry, or the rules governing allocation of heights and headings to aircraft. The most familiar and probably still the most common method of passing that information is to assemble the potential operators in a lecture hall and talk to them; or to assign them printed readings. Is that in fact the best way of guaranteeing successful action?

Secondly, there is the assessment of the operator once trained. We need to be sure that the training, however given, has truly imparted the necessary knowledge. The student may have been inattentive, or of too low ability, or a crucial part of the training may have been missed. Again, the most common form of assessment is verbal; in traditional British universities students write essays on broad topics to illustrate their mastery of the symbolic knowledge about those topics. In other teaching systems, they may answer large numbers of short questions, which can be scored more objectively than essays. A third possibility is a spoken interview, in which the interviewer can modify the questions to probe particular areas of knowledge about which the individual student seems uncertain. But all these are verbal methods of assessment. If it were true that a person can answer questions about a system, and yet fail to act correctly on the same system, none of these methods would be adequate.

Thirdly, most of the systems that are potentially hazardous seem clear cases for providing the operator with automatic assistance for carrying out portions of the task, for advising what actions should be taken in an unfamiliar situation, or for reminding the operator of some action that may have been forgotten. Such assistance can reduce the number of decisions that the operator has to take each minute, and compensate for failures of memory, training, or attention. But the knowledge that is embodied in an 'expert system' has to come from somewhere. Can the knowledge engineer extract it from a suitable human expert by talking to that person?

The intellectual problem

More generally, we can only guard against real-life errors by a theoretical understanding of the manner in which the human controller functions. It would not be safe to accept a chemical process merely because it seems to produce the correct product in one set of background conditions; to be sure that it will avoid instability in some other circumstances, we need to understand the underlying equations. In the same way the human component of the control loop needs to be analysed, if we are to be sure that a person who can carry out some emergency procedure on leaving the training school can also do so a year later, in the middle of the night, and in danger of death.

Examples of the kind of analysis we need have been provided by Rasmussen (1983) or by Anderson (1984). As the latter argues, the fact that humans can perform certain functions implies that they have within their heads some representation of general facts about the world, which we can think of as a set of propositions. This 'declarative' knowledge can be accessed in a variety of different situations when the task requires it, and can thus act as a kind of model of the world on which the decision-making system can try out the consequences of various actions. The system can also predict what will happen from seeing some event that it has not caused itself, and this will be true even though the event has not previously been experienced. Probably none of us have ever seen a reactor from which all the control rods have been removed, but most of us could make a rough prediction of the future!

It has been shown by Johnson-Laird (1984) that such a 'mental model' explains the results of a number of experiments on human handling of language and of reasoning, far better than various alternative theories that have been suggested. The notion of a mental model also fits very well with a distinction made by Rasmussen (1983). He discusses a 'symbolic' level of function that can be seen in task analysis of the performance of real operators. However, Rasmussen also distinguishes a 'rule-based' level of performance, in which the operator acts according to a classification of the present problem into some broad category. In this level of performance there may not be an exhaustive 'look-ahead' calculation of all future possibilities, and therefore, bizarre human errors may result.

In the same way, Anderson (1984) incorporates in his system a second kind of knowledge. In addition to facts about the world, a decision system needs to know the actions to take under certain conditions; without such knowledge the system would remain lost in thought. This 'procedural knowledge' can very naturally be represented as a set of condition–action rules, a 'production system' (Newell & Simon 1972), and in Anderson's formulation the two kinds of knowledge are represented in these two different ways. As Anderson points out, the logical distinction between the two kinds of knowledge does not mean that they are represented differently in the brain; computationally it would be quite possible for knowledge about appropriate actions to be held as a set of propositions rather than of rules. Equally, the propositional knowledge about the world could be held as a set of rules, and most expert systems intended for practical use do use such a representation. (The propositions are then implicit in the rules, but could be computed from them if necessary). It does however help to remind us of the logical distinction if, like Anderson, we think of different representations for declarative and for procedural knowledge.

At this stage we come back to the question of verbal knowledge about one's own actions. If all knowledge is represented in the nervous system in the same way, then we might expect it to be equally easy or difficult to elicit by questioning and in actual action. From a set of propositions that describe the functioning of a nuclear reactor, the operator could either derive the verbal answer to a question, or compute the correct action. If all knowledge were held in the same form, the lecture hall and the written examination would be good methods of teaching and assessing operators. If on the other hand there is some distinction of declarative and procedural knowledge, as Anderson supposes, there might also be greater difficulty in testing procedural knowledge by questions, or declarative knowledge by actions. The practical issues from which we started have come back to a theoretical issue about the form of representations in the brain.

2. The evidence

Signs of dissociation

Experiments on control

In a number of laboratory experiments, people have been practiced on some artificial control task in which the relations between each possible action and its consequences are known. In other experiments, people have been given verbal instruction about the relations and then asked to control with relatively limited practice. In either case, they can then be tested both for successful performance and for ability to answer objective-type questions about the relations. Cases have been found by Broadbent *et al.* (1986) and by Berry & Broadbent (1984), in which practice improves performance without improving success at question answering; and also in which verbal instruction improves question answering without improving performance. Funke & Muller (1988) also found that controlling a system (as opposed to merely watching events) improved performance without improving the ability to predict future events; and conversely, practice at predicting the future improved later predictions, but with no improvement in actual success at control. Both forms of training made the person less good at constructing a causal model of events within the system. Similarly, Berry & Broadbent (1984) found a negative correlation between the ability to control a system and the score obtained on questions about it; the people who could do one were to some extent less good at doing the other.

A criticism that may occur to some readers is that the particular questions chosen by the experimenter may not have been appropriate to the particular ideas of the person learning to control the system. They might have knowledge that is relevant to controlling the task, but which has not been tested by the questions. Stanley *et al.* (1989) asked people first to practice the task, and then to explain verbally to somebody else how to control it. Although in this case people could choose their own form of words, their own performance improved before they could tell somebody else how to succeed.

At a simple factual level, then, it is clear that people can sometimes be able to control a system without being able to talk about it, or to talk about it without being able to control it. The two ways of assessing knowledge are not totally separate, however. At a moderate level of practice they are dissociated, but more highly practiced 'expert' operators were found by Stanley *et al.* (1989) to be able to give verbal statements that would indeed help novices to perform more successfully. Even though a simple statement of the relations within the system may not help, and some kinds of true instruction may even be harmful (Berry & Broadbent 1988), yet it is possible to find certain kinds of verbal instructions that will improve performance (Berry & Broadbent 1988; Stanley *et al.* 1989).

It is particularly important that the presence of the dissociation may depend on the nature of the task. Berry & Broadbent (1988) found two tasks that were identical except for the presence of a lag in one of them, between each action and the appearance of the effect. The unlagged task was easier to learn, but with enough practice both tasks could be brought to the same level of performance. At that point, however, the operator could answer questions about the relations within the unlagged task, but not in the lagged one. There were also other differences (Hayes & Broadbent 1988); the lagged task could be disturbed by an unexpected reversal of the relations within the task, whereas the unlagged task was not. Again, the unlagged task was disturbed by asking the person to perform another task at the same time; but the lagged task was not. These facts suggest that the two, apparently similar, tasks are involving partly different mechanisms within the operator, as Anderson would suggest. That would be consistent also with another difference between the two versions of the task. The lagged task, but not the unlagged one, is disturbed by a verbal hint that might be expected to encourage the operator to try and learn the task in an 'explicit' and hypothesis-testing fashion rather than a more passive exploratory one (Berry & Broadbent 1988).

In the tasks that show the dissociation between verbal behaviour and correct decision, the key relations might be described as 'non-salient'. In those for which the operator's ability to perform agrees with their verbal knowledge, the variables are 'salient', obvious when the task is first approached. This can be illustrated by an experiment that required operators to control a simulated company, while at the same time interacting with a person representing the workers in the company (Berry & Broadbent 1987). Some of the relations in this system might well be expected; changes in the work-force affected the output of the company, and changes in the intimacy with which the person was treated had an effect on the intimacy shown in return. But there were also less obvious, hidden, relations; the degree of intimacy with the worker's representative also affected the output, and the size of the work-force affected the attitude of the worker's representative. Verbal questions about the salient and obvious relations revealed quite adequate knowledge, but questions about the non-salient relations were answered less well by practiced than by unpracticed people. Yet a test based on performance in the actual situation showed that the knowledge had in fact been acquired. Similarly, Porter (1986, 1988) asked people to control a task with two aspects; one was a straightforward sensori-motor reaction, with few and obvious variables. The other was a complex handling of strategic decisions. In the first, the verbal statements of the people agreed well with objective performance; in the second, they did not. Correspondingly, the first task was badly disturbed by a simultaneous other task, while the second was not.

Experiments on tasks other than control

While these developments have been taking place in the study of control, similar observations have come to light by using different tasks. Reber (1967) (Reber *et al.* 1980) examined the learning of artificial grammars; the person is required to learn sequences of letters, which are in fact all governed by abstract rules that allow only some kinds of sequence. After this experience, the person is able to discriminate new test sequences into grammatical and ungrammatical, but cannot give a coherent verbal account of the rules. It is important that the person does not know that there are rules when experiencing the sequences, but merely tries to learn the individual items; deliberate instructions to try and learn rules may actually slow down learning (Reber *et al.* 1980). This is suggestively parallel to some of the results of Berry & Broadbent (1988).

A second line of attack comes from memory experiments; in such experiments people may show effects of past experience by tests such as ease of identifying words in a brief flash, or judging how much they like one word rather than another. Yet they are unable to recall or even to recognize that the words have been seen previously (for a review, see Schacter (1987)). The different kinds of task may be affected differently by brain injury.

A third group of experiments involves 'concept learning', for example, Lewicki (1986). People are asked, for example, to look at photographs of young women and to judge their intelligence. Before doing so, the judges have previously seen a sample of similar pictures together with verbal descriptions of the women; for some judges, women with short hair were described as more intelligent, while for other judges, women with long hair were described as intelligent. The judgements made on a fresh set of pictures reflected the particular experience the individual judge had been given, yet the judges were unable to identify that they were being influenced by hair length.

There are therefore several lines of evidence showing that the verbal accounts people give of their mental processes may be out of agreement with their actual performance.

3. Theoretical implications

The evidence that has been reviewed is reasonably consistent with Anderson's distinction between declarative and procedural knowledge, as involving different mechanisms or representations in the human decision system. Although consistent with it, however, some of the evidence does not exclude alternative interpretations.

Accessibility of memory by various routes

It has to be recognized that this area is controversial; in everyday life we know very well that somebody who understands a situation can tell us about it and also take appropriate actions. If they are wrong in their actions, we are not likely to trust anything they may say to us about the situation. It seems to run counter to this commonsense view if we suppose that people can show knowledge in one way and yet not in another. Furthermore, admitting that such a thing can happen seems to require two decision mechanisms, or two kinds of knowledge, within the brain. Perhaps it also needs some kind of barrier or insulation between them. That is a complication that most scientists would prefer to avoid if they possibly can.

Admittedly, we have seen plenty of experiments in which the things people say do not correspond to the things they do; but the weak point is that correct action may be based on a belief that is not the one the experimenter has asked about. It may indeed be based on a false belief. If I believe my neighbours are spraying my lettuces with arsenic, that can be false, and yet the fact that it makes me wash each salad before I eat it may be good for my health. Perhaps some similar fantasy underlies action in each of the experiments I have mentioned.

Logically, one cannot convince a sceptic about this, because it is impossible to ask questions about every conceivable relation that might exist. The best one can do is to make such a view implausible, by asking questions directly about the same measures that are taken from performance (Broadbent *et al.* 1986), by trying a variety of different kinds of questions (Berry & Broadbent 1984), or by leaving the person free to express their own beliefs whatever they may be (Stanley *et al.* 1989). It is always possible that some other measure of verbal knowledge would agree with action; one cannot prove a negative.

In addition, it is clear from the evidence that different measures of verbal knowledge are not equivalent. From general knowledge about human memory, we could not expect that they would be. If we show a person a list of words and then test memory, different tests will reveal different amounts of knowledge. Typically, a simple request for the person to recall all the words will give only a limited degree of success. Many of the words that have been 'forgotten' will however be recognized correctly if they are shown mixed with others (for example, Shepard 1967). They may even be recalled if some hint associated with the correct word is given (Tulving & Pearlstone 1966). Although information is in the brain, it is not accessible to one kind of test even though it can be shown through another kind of test.

To some extent this is a matter of one test being more sensitive than another, but that is not the whole story. If the context is deliberately varied, it is possible to make a recognition test fail for items that the person can recall (Tulving & Thomson 1973; Tulving 1983). Under special circumstances, people will recognize items they cannot recall just as well as items they can; the two tests are independent, rather than one being more sensitive than the other (Broadbent & Broadbent 1975, 1977; Rabinowitz *et al.* 1977). The key factor is that the situation should provide the correct context when memory is interrogated, so that the information about the original event can be retrieved. Just as a computer database may find it much easier to access a record when probed with a descriptor assigned to that record when originally filed, so people may also remember something in one context and forget it in another.

Such mechanisms are undoubtedly active in some of the situations usually described as studies of implicit and explicit knowledge. Thus, for example, Dulany *et al.* (1984) used Reber's technique of learning artificial grammars, but tested for explicit knowledge in a different way. Instead of asking merely for statements of the rules, they presented sequences of letters and asked people to show which part of the sequence was guiding their decision. This method revealed some knowledge of the regularities in the strings, that may have been enough to control the decisions about grammaticality. Similarly, in the control task of Berry & Broadbent (1987), the method used to reveal knowledge of non-salient relationships between personal relationships in the factory, and output of the factory, was to put the person under test back in a sample of the task rather than to give questions on paper.

Perhaps particularly striking is a result of Marescaux *et al.* (1989) using the same control task as that employed by Berry & Broadbent (1984). After people had practiced the task and could control it, they were asked 'questions' about the task in the form of a series of problems shown to them at the computer, so that they saw just what they might have seen in the full task. In that case, some knowledge of the task was revealed, but only on some of the questions. The key factor appears to be that the person was tested on specific situations which they had themselves experienced while practicing. They did not seem to have learned something that could be used in other novel situations.

It is therefore clear that knowledge may be elicitable by tests that recreate the original context, even though it may fail to appear on tests that differ too much from the situation during learning. Some of the studies mentioned in the last section do not explore this possibility. For example, the work of Reber or of Lewicki concentrates on the discrepancy between performance and the ability to state principles, without going into the degree of knowledge that might appear in situations resembling those of original learning. For this reason, one can only draw limited conclusions from the experiments that show discrepancies between verbal behaviour and action. They do not make it necessary to accept Anderson's distinction of declarative and procedural knowledge. Action may be based on some untested form of knowledge, and the original context may fit better with the measure of action than it does with the measure of verbal knowledge. Again, in other situations verbal knowledge may be more accessible than correct action is. It is also possible that different tests of verbal knowledge may disagree with each other, and that some of them will agree with action while others do not.

Modes of learning and forms of representation

Merely demonstrating a dissociation then will not convince anybody that there are fundamentally different mechanisms for implicit and for explicit knowledge. If indeed the mechanisms are different, however, they should be affected by different experimental variables. It should in that case be possible to find tasks in which action and verbal report appear to gain access to the same knowledge, and closely similar tasks in which they do not. By looking at the exact differences between such tasks we may shed light on the difference between mechanisms. This is a much more hopeful strategy for the future than merely demonstrating or denying dissociation.

So far, perhaps the clearest case of such a comparison is the pair of tasks devised by Dianne Berry and described above (Berry & Broadbent 1988). In both tasks, each control action by the person produces an output from the system that depends only on that input and simply adds a constant to it. In the 'salient' case, the output appears immediately; and when people have learned to control this version they can answer questions about it. In the other, 'non-salient', version, the output appears only after the next input, so that each output is undetermined by the last input. This version shows the dissociation of action and performance, and as noted earlier, the two tasks show a number of other differences. Thus the two tasks do seem to be learned in some rather different way.

Although these two tasks are the most nearly identical ones in the literature, the differences between them are consistent with a number of other findings. First, as has been noted above, Porter (1986, 1988) observed two, fairly different, tasks within a single general situation. The task on which verbal knowledge was better was also the one more disrupted by a secondary task. Secondly, the concept of salience was derived from a study by Reber *et al.* (1980), in which strings of letters were shown to the person either grouped according to the grammatical rule they illustrated, or mixed together. The first form of presentation showed more beneficial effects of explicit instructions to learn the rules, just as the salient task of Berry & Broadbent did. Thirdly, in the tasks of Broadbent *et al.* (1986), practice of a single relations on its own improved verbal knowledge as well as performance, whereas practice of multiple relations together did not; and experience of a complex crossed relation caused some deterioration of verbal knowledge when a simpler version of the same relation did not.

It is worth mentioning also the suggestive parallel between the concept of 'salience' in the Anglo-Saxon literature, and that of 'transparency' in German; the interest in that case has not been so much concerned with the relation of performance to verbal report, as with the relation to tests of intelligence. Studies of system control have been found several times to show little relation to intelligence, as long as there are many variables and the relations between these variables are concealed. When however, key relations are available to the operator, then test intelligence becomes more relevant. (For reviews in English, see Putz-Osterloh & Lemme (1987); Funke (1988); Funke (1990).)

The tasks that show dissociation seem therefore to be those in which the person has to learn to act in a situation, and yet the key features of that situation are unclear because they are accompanied by many other events. Even if the key features have been explained previously to the person, they may not be recalled at the correct moment in the task unless the person is deliberately asked to say them at that time (Berry & Broadbent 1984). Hence the learning takes the form of building links between situations and actions, 'procedures' in Anderson's sense. Such learning is best built up unselectively by exposure to the task itself.

Tasks that show good agreement between action and speech, on the other hand, are those with few variables and with clear relations between them, which can reasonably be learned by selective concentration on those few variables and by testing hypotheses about possible relationships. The resulting knowledge can most simply be described, not as links between situations and actions, but as a set of propositions about the underlying variables; if the proposition is known, it may produce an entirely new correct action if the situation is new. Propositional knowledge however corresponds more to Anderson's declarative than to his procedural knowledge.

The strategy or mode of learning that is going to be most successful is likely to be different for the two kinds of task, so that for a salient or transparent task, learning will be better with verbal instruction and will transfer easily to new situations; for a non-salient task, learning will be better through experience, badly tested by questioning, and transfer only weakly to new situations. If experiment shows that the best conditions for learning reverse when the task is different, that is much stronger evidence for the existence of two mechanisms than is given by the mere demonstration that people can perform without being able to talk about it. Although there are few reversals of relations in the literature, we have seen some.

Remember for example, that Berry & Broadbent (1984) found, across their whole body of experiments, a significant negative correlation between the ability to perform well and the ability to answer questions about the situation. The people who were best at one were actually worse at the other. Similarly, Funke & Muller (1988) found that factors improving scores by one measure gave significantly negative results on another measure.

It should not be surprising to find negative relations because in machine intelligence it is found that the order of merit of various computational strategies may reverse from one situation to another. From declarative knowledge of the possible moves in chess, a chess-playing program can compute possible future consequences for any action and therefore do much better than a human amateur who fails to notice some possible trap. But in certain situations, such as the end-game, such a program may be handicapped by the very large number of alternatives. It may then lose to another program (or even a human) that operates on procedural knowledge, choosing an action by consulting a previously calculated table that lists actions appropriate to situations. On that strategy, the results of all the other possible moves are never calculated. The kind of tasks that are best for calculation from propositional knowledge may well be worst for the application of condition–action rules, and vice versa. Any realistic chess program uses both techniques as appropriate; and it is reasonable that human beings should do the same (Michie & Johnston 1984).

Conclusions

At the very least, we have to recognize that the conditions of testing knowledge are vitally important in deciding whether it will be revealed. The evidence for that is overwhelming. Even if all knowledge is ultimately represented in the same form, therefore, access to it is by different paths.

Although most of the evidence merely shows this more restricted point, there are a few studies that go further than that. The comparisons of different tasks do establish that a distinction of procedural, condition–action, knowledge from declarative, event–event knowledge, is functionally useful. Some kinds of representation are useful only if the conditions of performance are very similar to those of learning; others are more portable from one situation to another. The ultimate basis in the nervous system may well be the same for both types, much as neurones operate on the same basis in many different parts of the brain. Just as it is useful to distinguish the auditory from the visual pathways, however, without implying a different kind of neurone, so it is useful to keep declarative and procedural knowledge separate in our thinking.

4. Practical implications

If we return now to the practical questions raised in the introduction, we have seen evidence that supports the views of many non-academics. It is not safe to rely on classroom instruction for the training of those who take decisions in complex systems. Equally, one cannot always assess the knowledge of such people by, for example, a written test or a promotion interview. If in despair one decides to build into a non-human system the knowledge possessed by the human expert, it will not necessarily be possible for the expert to reveal that knowledge to you.

However, the evidence also points to a real role for knowledge that is verbal, propositional, or declarative. That type of knowledge is superior when the situation is transparent; it is useful in turning apparently opaque tasks into transparent ones by isolating the correct key variables. It can be used to train operators to call up relevant knowledge at the right moment in the task. Above all it allows the person who possesses it to react in novel situations when those who are basing action on intuitive procedures may be adrift.

The true lesson therefore is not that intuition should always triumph, and that in risky situations you should automatically let 'the force be with you'. Rather, we need to design tasks and training situations so that the computational advantages of each way of operating are exploited. To do this will need explicit statements about the scope and limits of implicit knowledge.

The author is employed by the Medical Research Council. The experiments of Dr Dianne Berry were supported by the Economic and Social Research Council.

References

Anderson, J. 1984 *The architecture of cognition*. Cambridge, Massachusetts: Harvard University Press.

Berry, D. C. & Broadbent, D. E. 1984 On the relationship between task performance and associated verbalizable knowledge. *Q. J. exp. Psychol.* **36**A, 209–231.

Berry, D. & Broadbent, D. E. 1987 The combination of explicit and implicit learning processes in task control. *Psychol. Res.* **49**, 7–15.

Berry, D. & Broadbent, D. E. 1988 Interactive tasks and the implicit–explicit distinction. *Q. Jl exp. Psychol.* **36**A, 209–231.

Broadbent, D. E. & Broadbent, M. H. P. 1975 The recognition of words that cannot be recalled. In *Attention & performance V* (ed. P. M. A. Rabbitt & S. Dornic), pp. 575–590. New York: Academic Press.

Broadbent, D. E. & Broadbent, M. H. P. 1977 Effects of recognition on subsequent recall. *J. exp. Psychol.: general* **106**, 330–335.

Broadbent, D. E., FitzGerald, P. & Broadbent, M. H. P. 1986 Implicit and explicit knowledge in the control of complex systems. *Br. J. Psychol.* **77**, 33–50.

Dulany, D. E., Carlson, R. A. & Dewey, G. I. 1984 A case of syntactical learning and judgment: how conscious and how abstract? *J. exp. Psychol.: general* **113**, 541–555.

Funke, J. 1988 Computer-simulated scenarios: a review of studies in the F.R.G. *Simul. Games* **19**, 277–303.

Funke, J. 1990 Solving complex problems: human identification and control of complex systems. In *Complex problem solving: principles and mechanisms* (ed. R. J. Sternberg & P. French). Hillsdale, New Jersey: Erlbaum.

Funke, J. & Muller, H. 1988 Eingreifen und prognostizieren als determinanten von systemidentifikation und systemsteuerung. *Sprache kog.* **7**, 176–186.

Hayes, N. & Broadbent, D. E. 1988 Two modes of learning for interactive tasks. *Cognition* **28**, 249–276.

Johnson-Laird, P. N. 1983 *Mental models*. Cambridge University Press.

Lewicki, P. 1986 *Nonconscious social information processing*. Orlando, Florida: Academic Press.

Marescaux, P.-J., Luc, F. & Karnas, G. 1989 Modes d'apprentissage selectif et non-selectif et connaissances acquises au controle d'un processus: evaluation d'un modele simule. *Cah. Psychol. Cog.* **9**, 239–264.

Michie, D. & Johnston, R. 1984 *The creative computer: machine intelligence and human knowledge*. Harmondsworth, Middlesex: Penguin.

Newell, A. & Simon, H. A. 1972 *Human problem solving*. Englewood Cliffs, New Jersey: Prentice-Hall.

Porter, D. B. 1986 *A functional examination of intermediate cognitive processes*. D.Phil thesis, University of Oxford.

Porter, D. B. 1988 Computer games and human performance. In *Proceedings, 11th Symposium on Psychology in the Department of Defense*, pp. 251–255. Colorado Springs: United States Airforce Academy.

Putz-Osterloh, W. & Lemme, M. 1987 Knowledge and its intelligent application to problem solving. *Ger. J. Psychol.* **11**, 286–303.

Rabinowitz, J. C., Mandler, G. & Patterson, K. E. 1977 Determinants of recognition and recall: accessibility and generation. *J. exp. Psychol.: general* **106**, 302–329.

Rasmussen, J. R. 1983 Skills, rules, and knowledge; signals, signs, and symbols, and other distinctions in human performance models. *IEEE Trans. Syst. Man Cybern.* **13**, 257–266.

Reber, A. S. 1967 Implicit learning of artificial grammars. *J. Verbal Learn. & Verbal Behav.* **5**, 855–863.

Reber, A. S., Kassin, S. M., Lewis, S. & Cantor, G. 1980 On the relationship between implicit and explicit modes of learning a complex rule structure. *J. exp. Psychol.* **6**, 492–502.

Schacter, D. L. 1987 Implicit memory: history and current status. *J. exp. Psychol.* **13**, 501–518.

Shepard, R. N. 1967 Recognition memory for words, sentences, and pictures. *J. Verbal Learn. Verbal Behav.* **6**, 136–143.

Stanley, W. B., Mathews, R. C., Buss, R. R., & Kotler-Cope, S. 1989 Insight without awareness: on the interaction of verbalization, instruction and practice in a simulated process control task. *Q. Jl exp. Psychol.* **41**A, 553–577.

Tulving, E. 1983 *Elements of episodic memory*. Oxford University Press.

Tulving, E. & Pearlstone, Z. 1966 Availability versus accessibility of information in memory for words. *J. verbal Learn. verbal Behav.* **5**, 381–391.

Tulving, E. & Thomson, D. M. 1973 Encoding specificity and retrieval processes in episodic memory. *Psychol. Rev.* **80**, 352–373.

Phil. Trans. R. Soc. Lond. B **327**, 503–512 (1990)

Printed in Great Britain

Human error on the flight deck

By R. Green

Royal Air Force, Institute of Aviation Medicine, Farnborough, Hants, U.K.

Despite terrorist bombs and structural failures, human error on the flight deck continues to account for the majority of aircraft accidents. The Royal Air Force (RAF) Institute of Aviation Medicine (IAM) has investigated the psychology of such error since the early 1970s, and to this end has used two principal techniques. The first has involved assisting in the official inquiries into both RAF and civil flying accidents, and the second has involved setting up a reporting system that permits any commercial pilot to report his own everyday errors, in complete confidence, to the RAF IAM. The latter system possesses the clear benefit of gathering error data untainted by considerations of culpability, and sometimes permits system rectification before the occurrence of accidents. This paper examines selected examples of errors associated with the design of equipment and with the social psychology of crews, and suggests that some consideration of the psychology of organizations may be necessary to ensure that the problems of human error are given the degree of consideration they require.

Introduction

It has become cliched for those writing about aircraft accidents to point out that flying is, compared with other forms of transport, safe. Commercial jet transport aircraft are lost at a rate of about one per million flying hours, so there is some degree of contradiction in including the consideration of such events in a symposium on hazardous situations. The individual perception of 'risk' is determined, however, not just by the probability of a certain outcome, but by the subjective utility of that outcome. The negative utility of being involved in an aircraft accident is enormous, and this naturally affects the importance of the subject to the passenger. It is also true that when an aircraft accident occurs, it is such a large and public event that it cannot be ignored, and this may help to generate the measure of perhaps irrational anxiety about flying that undoubtedly exists in a significant proportion of the population.

The attention consequently focused on flying accidents has meant that errors in aviation have been investigated more thoroughly than errors in any other sort of endeavour, and the solutions that have been put in hand may well have lessons for those in less well-researched disciplines. What then, makes flying safe? A common belief is that the commercial pilot is a singular individual, carefully selected for his or her high degree of astuteness and particular aptitude for the job. In fact, the average holder of a commercial flying licence in the U.K. has a performance on a test of adult intelligence about equal to that of the average student at a teacher training college, and rather lower than the average undergraduate's. The range of intelligence scores achieved by pilots is very wide, yet even the poor performers have been able to demonstrate sufficient competence at flying to gain a commercial licence. If safety does not appear to stem from the intrinsic quality of the pilots, what is its source?

There are probably two main factors safeguarding flying from human error. The first, and probably most important, is that commercial flying has become extremely regulated and 'proceduralized'. In flying there is a 'procedure' for every predicted eventuality. In leaving any airport the pilot will be provided with a set of standard departure patterns that define the routes and heights that he must achieve during that departure, and the same goes for arriving at an airport. If an engine catches fire, the pilot will not need to invent or think through the best course of action: it will be written down in his flight reference cards. In starting or shutting down the aircraft, the safest procedure will have been worked out and embodied in a drill. This process has meant that everything possible in flying has been reduced (in terms of one set of jargon) to a 'rule-based' activity (Sanderson & Harwood 1988). High-level decisions are made as infrequently as possible on the flight deck as every contemplable set of circumstances will have been discussed, and the best solution and procedure decided upon in advance. This has not always been so in flying, however, since there was natural resistance among pilots to see every aspect of their job reduced to the exercise of some predetermined set of responses, leaving much less within their immediate locus of control. The pressure for standardization has come from safety considerations, and O'Connor (1987) has pointed out that in 1933, when flying in the U.S.A. was regulated in a way that did not happen in the U.K. until the 1950s, U.S. airlines flew 21686515 passenger miles per fatality, whereas the British figure was 1080000.

The second major reason why the system is safe is the emphasis that is placed on the training and competency checking of airline pilots. The pilot must not only demonstrate his general competence at flying before gaining a commercial licence, but he must hold specific endorsements on his licence for each aircraft type that he flies. Even then, he must pass regular checks in the simulator and when flying on the line so as to remain legal, and he must be retested if his level of flying activity (or skill maintenance) drops below certain prescribed minima. It is probably true to say that in no other profession in which errors can threaten life, such as medicine or air traffic control, is the maintenance and checking of competence so thoroughly addressed.

The two factors considered above address exclusively human factors considerations: minimizing the scope for error in flight-deck decision making through proceduralization, and ensuring that pilots are competent at exercising the procedures for their aircraft. Neither of these factors has required the application of any particular psychological expertise as the requirements and solutions have been largely obvious. Unfortunately, serviceable aircraft continue to crash, and the remainder of this paper is concerned with the human factors that remain relatively unaddressed in flying. The data for the conclusions that are drawn come from two main sources. The first is the attendance of psychologists from the Royal Air Force (RAF) Institute of Aviation Medicine (IAM) at the inquiries into (and interviews with the crews involved in) both military and civil flying accidents. The second data source is a scheme known as the Confidential Human Factors Incident Reporting Programme (CHIRP)†. This enables all civil airmen and air traffic controllers to report their errors, not anonymously, but in complete confidence, to the RAF IAM. Each year about 200 pilots and air traffic controllers take the opportunity to use this scheme to tell us about the mistakes they have made in the air and why they believe they made them. No attempt is made here to give a comprehensive account of the human factors that cause incidents and accidents but only to provide a set of examples that typify certain problem areas.

† The Confidential Human Incident Reporting Programme was initiated and is sponsored by the Civil Aviation Authority and is operated by the RAF Institute of Aviation Medicine, Farnborough, Hants, U.K. Information on the scheme and its publication, *Feedback*, may be obtained from this Institute.

Hardware

The first area that is always assumed to attract a great deal of attention in aircraft development is that of the design of the controls and displays used by the pilot, and this is, at least to some extent, true. Unfortunately cockpit designs are constrained by cost, space, and the number of operators that have been decided upon. There are also many problems that can arise through the life of an aircraft as it is modified to meet new requirements that may render an initially adequate arrangement less acceptable. The following report was submitted to the CHIRP system by a helicopter pilot, and concerns the way in which pressure settings were demanded and displayed on his helicopter's altimeters. 'At 2000 ft my co-pilot said, "You've gone below 2000 ft!". I replied that I had not, but then saw that my altimeter was set on 1030 mb and not the correct QFE (airfield ground level pressure setting) of 1020 mb'. Consider the altimeters shown in figure 1. The altimeters are viewed from a distance of some 50 cm, while the instrument panel is acknowledged to suffer from shake. In the AS332 fleet of aircraft, the individual helicopters are fitted with altimeters of types (*a*) and (*c*) or (*b*) and (*c*). As the pilots fly from either seat, according to crewing requirements and convenience, a pilot may find himself using an instrument of type (*a*), (*b*) or (*c*).

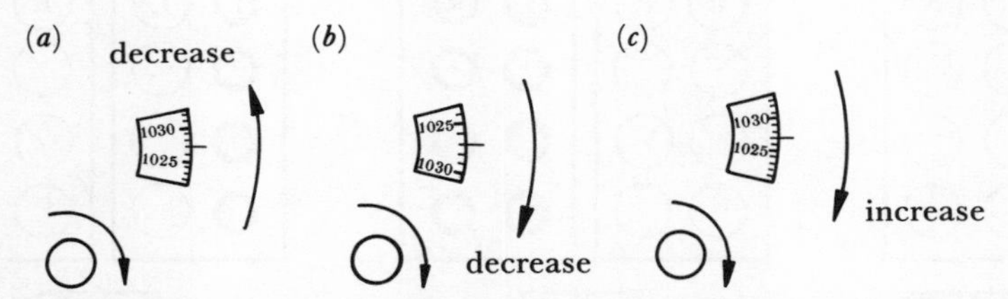

Figure 1. Arrangements of barometric setting controls and displays on helicopter altimeters.

These altimeters are superficially similar, but the sub-scales and the mode of changing the datum pressure setting are all different. It seems that most of my colleagues have difficulty in seeing and setting the correct pressures. Whatever happened to the altimeters with veeder counters for the pressure setting that we used to have 20 years ago?

It requires no knowledge of ergonomics of man–machine interface design to see what is wrong with the altimeters described above, or to imagine the nature of the accident that they could bring about since a wrongly set altimeter can easily cause ground impact. The practical decision that has to be made, however, is whether these altimeters are so unsatisfactory that the regulatory authority should compel the operating companies to go to the expense and trouble of replacing or standardizing them. The altimeters clearly function, and who is to assign a probability to their causing an accident during their operational life? Had the altimeters actually caused an accident, however, they would obviously be changed since we would know that the above probability is 1, and it is this fact which determines that flight safety is a process driven far more powerfully by failures that result in accidents than by identified system shortcomings. An airline operator has justified this situation as follows:

> Aircrew should be expected to be reasonably intelligent alert human beings who are able to assimilate that they are liable to normal human error. Consequently, they should be prepared to accept these errors are their own responsibility and not palm everything off on some designer or management who expect a fair day's work for a fairly generous salary. A crew member who by his own admission is previously aware that the switch positions are reversed on two similar aircraft is surely capable of considering mis-selection as soon as a problem appears.

It has been implied that although ergonomic problems may be powerful progenitors of human error, they do not demand intellectual solution but simply an appropriate appreciation of the balance between risk and cost on the part of operators and regulators. It is clearly part of the task of the applied psychologist to evaluate the risk that may be inherent in a piece of design, but broader interests will inevitably need to be taken into account, and the role of the psychologist may be less clear, in evaluating the balance between the probability of hazardous failure and the economic cost of rectification.

Unfortunately, not all ergonomic problems on flight decks are as amenable to solution as that described above. A problem of current interest concerns engine instrumentation. Such instruments may be divided into those required for setting engine parameters and those providing status information such as oil temperature. The 'setting' instruments have traditionally been located on the front panel of the cockpit, arranged in columns aligned with the appropriate throttles and in rows of identical instruments (see figure 2*a*). The 'status' instruments may, on a three-man flight deck, be displayed on a panel behind the pilots as it will be the third pilot or engineer whose task it is to monitor them.

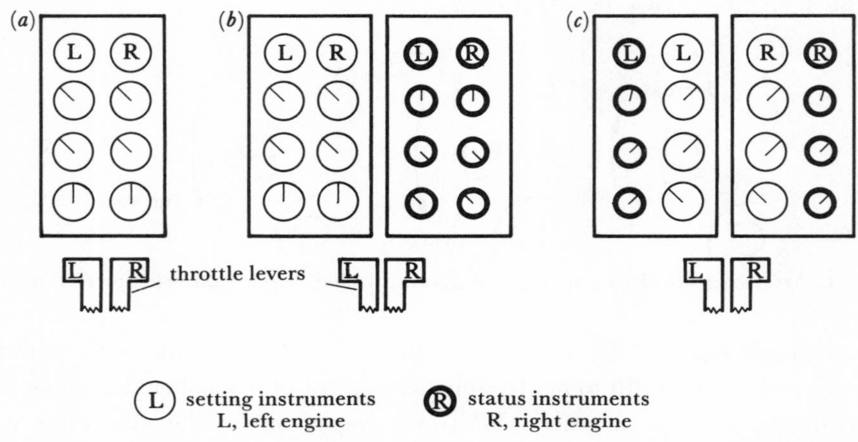

FIGURE 2. Possible arrangements of engine instruments on twin-engine aircraft.

On a two-man flight deck, however, all of the engine instruments must be accommodated in view of the two pilots. Since the front panel is limited in height, is it better to have arrangement (*b*) shown in figure 2, in which the status instruments have displaced the setting instruments to the left, or is it better to leave the setting instruments aligned with their respective throttle levers and split the status instruments so that they are placed on either side of the setting instruments (arrangement (*c*)?) (*c*) Has the advantage that the instruments for each engine are kept together and in line with the landmarks provided by the throttles, but B has the advantage of keeping similar instruments together and making deviant readings easier to identify. In fact, most two-pilot aircraft have arrangement (*b*), but a recent accident in which the crew shut down the serviceable rather than the faulty engine will doubtless cause some attention to be given to the wisdom of the practice.

Aircraft equipment today is far more flexible and capable of being more closely tailored to the human's requirements than ever before, and this means that the onus has moved from training the pilot to cope with what is practically achievable to designing a system that matches the human's abilities. A specific example of this process is provided by the display of attitude (pitch and roll) information to pilots. Traditional attitude indicators (AIs) appear as they do because the original devices consisted of a vertical gyro with a horizon card and aircraft symbol attached to its front. Although the attitude indicator in a modern aircraft is likely to derive its data not from a gyro but from a laser inertial platform, the image on its electronic display would be familiar to the users of the very earliest gyro instruments.

Such consistency is obviously advantageous in minimizing the likelihood of negative transfer of pilot skill between one aircraft and another, but pilot disorientation continues to be a lethal problem in flying, and conventional AIs may have a lot to answer for. They provide information to the central or focal part of the visual system that provides the dominant visual input to conscious attention, but do not drive the peripheral retina that is so important in detecting the movements of the world, vection cues, which are probably essential for providing a normal perception of orientation. Traditional AIs therefore compel instrument flying to be a conscious and rather complex skill rather than an exploitation of natural human abilities. Attempts are now being made by life scientists (Green & Farmer 1985) to develop wide field of view attitude indicators to match the machine to the man instead of vice versa.

The balance between the work that is done by man and that which is done by machines on the flight deck has progressively shifted since the first jet airliners were developed. On these it was common for there to be two pilots, a navigator, a flight engineer, and a radio operator. Automated engine management and navigation have caused all but the two pilots to be dispensed with, and automation in military aircraft has reached the stage at which the term 'electronic crew member' has become part of the accepted jargon (Taylor 1988). How to integrate this electronic individual with the other crew members is more than a simple design problem, as the manner in which the pilot models his environment may be changing in a fundamental way.

On older aircraft the pilot knew that the airspeed indicator was driven by the pitot head sensor in a way that he could comprehend. By integrating information from various simple displays like this, the pilot was able to generate his model of what the aircraft was doing in space. Today, however, information is sensed in highly sophisticated ways and combined by the aircraft's computers to be presented on an integrated electronic display. Such displays are generally reliable, well engineered, easy to interpret, and a pleasure to use. The potential problem is that as they are so seductive, and since the pilot cannot possibly understand the technology involved in the generation of the display, he is compelled to use the display itself as his world model rather than creating his own internal model from the raw data. Although this shift in modelling mechanisms is likely, overall, to reduce error, it may generate a different class of problem in which the pilot, used to trusting his technology, actually trusts it too much and fails to utilize cues that should suggest to him that an error has crept in. The most spectacular example of his form of error that has so far occurred concerned the Air New Zealand DC10 that collided with Mount Erebus in Antarctica. The aircraft's inertial navigation system's computers had been programmed with an incorrect position, but the crew trusted them and were probably seduced into interpreting external visual information in a way that conformed with the world model generated for them by the aircraft.

Although equipment design, or traditional ergonomics, is not a topic that is presently in vogue among those who theorise on human error, the purpose of the foregoing has been to emphasize that these micro-factors in the aviation system are still matters of critical importance that must constantly and individually be addressed by those human factors specialists who are engaged in bringing about practical improvements to safety.

Social Factors

Equipment deficiencies have their effects, by and large, on the individual operator, and there are many perceptual and skill problems of flying not addressed here of which the same could be said. The individuals on a flight deck are required, however, to operate as a team or social group and it is only relatively recently that this flight deck team has been studied to identify the ways in which their interactions may affect operational safety. Such a study suggests that many of the concepts, such as conformity and compliance, already existent within social psychology are adequate to describe the events observed in reported breakdowns of team behaviour. The idea of 'risky shift' in group decision making (that a group is likely to make a more risky decision than the average member) is clearly illustrated in the following CHIRP report from the first officer of an aircraft that had already been forced to overshoot from two approaches in poor weather.

> A suggestion was made that we should fly down 50 feet lower (than the decision height – the height at which the runway must be visible for the approach to continue) but as this was not legal, it was ruled out. We then managed to delude ourselves that flying level at the decision height of 220 feet was a legal and reasonable way to achieve our landing... From my position I studied the information on the flight instruments with a growing feeling of unease... We touched heavily on the centre line. Heavy braking followed what can only be considered a 'max performance' stop... The pregnant silence which followed served to reinforce our feelings that we'd been party to an act of supreme folly and bravado and were lucky to escape with a few grey hairs and severely battered pride.

Perhaps a more common problem in the flight deck team is that associated with the way in which leadership is exercised by the captain and the influence that this has on the propensity of the junior members of the crew to question his or her decisions. Examples of such problems are quoted by both Wheale (1983) and Foushee (1984), and the two which follow are from the transatlantic analogue of CHIRP. The first concerns a captain who was ignoring a speed restriction from air traffic control. After the co-pilot had attempted a number of times to bring the captain's attention to the restriction, the captain's response was 'I'll do what I want'. Air traffic control issued further requests which the copilot attempted (unsuccessfully) to persuade the captain to comply with, until eventually the captain told the co-pilot 'just to look out the damn window'. Breakdowns in coordinated behaviour are not always associated with such obvious dominance on the part of one crew member, however, as the second example makes clear.

> The captain said he had misread his altimeter and thought he was 100 ft lower than he was. I believe the main factor involved here was my reluctance to correct the captain. This captain is very 'approachable' and I had no real reason to hold back. It is just a bad habit that I think a lot of co-pilots have of double-checking everything before we say anything to the captain.

Even when danger is clearly imminent there can still be reluctance to take assertive action. In a recent helicopter accident in the North Sea, the aircraft was manned by two captains, but the handling pilot suffered a temporary incapacitation as he came to the hover before landing on an oil rig. The aircraft descended rapidly, and the cockpit voice recorder shows that although the non-handling pilot appreciated that matters were extremely abnormal, he initially tried to gain the other pilot's attention by asking him what the matter was, and actually took control too late to prevent the aircraft from hitting the sea.

The complexities of the relationship between the pilots on the flight deck are well illustrated by the incident which occurred to a BAC 1-11 aircraft landing at Gatwick on 12 April 1988 (AAIB 1989). The main runway was out of commission and aircraft were landing on runway 08L. This was known as the emergency runway and normally acted as the parallel taxiway for the main runway. When in use as a runway, 08L was lit with good quality edge lighting, but when used as a taxiway it was lit by a set of green centreline lights. There is a further parallel taxiway to the left of 08L and this was also illuminated with green centreline lights. The result of this rather complex description is that when the main runway was in use the pilot was presented with a visual picture of an obvious runway with a taxiway to its left, and when the emergency runway was in use the visual picture was very similar. As many pilots were aware that the emergency runway was 'the taxiway to the left of the main runway', there was a danger that, during main runway closure, a crew might believe the emergency runway to be the main runway and consequently land on the most northerly taxiway believing it to be the emergency runway.

The aircraft that actually landed on this taxiway contained a two-man crew. In the left seat (normally the captain's) was a very steady and unhurried individual who was actually the legal first officer, but who was acting as the captain on his last check trip before being made a full captain. He had given the handling of the aircraft to the man in the right seat who was a rather more assertive sort of individual, was a training captain, was the legal commander of the aircraft, was checking the aspiring captain, but was behaving (for the purposes of the check) as though he were the first officer or co-pilot. The latter individual had aligned the aircraft with the correct, emergency, runway and the approach was progressing normally. The first officer was unhappy with the visual scene, however, and wanted to confirm with the captain that both of them were clear what they were doing. He consequently said to the captain words to the effect of 'You are going for the emergency runway aren't you?'. The effect of this remark on the senior, handling pilot was instantly to make him believe, inaccurately, that he was flying towards the main runway and that he should be flying towards the taxiway to the left of the main runway. Since he did not wish the first officer to think that he had made such an error he replied 'Yes, of course I am', and then surreptitiously changed course for the taxiway while the first officer's attention was directed inside the cockpit. The first officer did not look up until too late to prevent the aircraft from landing on the taxiway and narrowly avoiding another aircraft that was taxiing towards it.

The crew members of this aircraft were mature and competent individuals who were not tired, and who had a friendly and natural relationship with one another, but who nevertheless did not wish to be as frank with one another as system safety dictated. Study of incidents such as these suggests that the four main factors that determine how crew members will interact are their personalities, their statuses, their roles, and their relative abilities. It is obvious that if personality and status act in the same direction (that is, a dominant captain is paired with a submissive first officer), then there is a considerable likelihood that events will have to be serious before the first officer will choose to challenge the captain. Slightly more surprising perhaps, are the examples that show how hazardous the situation has to become before the pilot with the non-handling role will intervene in the behaviour of the handling pilot, especially if he believes the handling pilot to be competent. The Gatwick landing incident involved the interaction of all these factors, but it also required the potential for failure provided by the ambiguous runway lighting.

To prevent the recurrence of such an incident it is clear that modification was required to the runway lighting, but incidents such as this also call for a programme of behaviour modification on the flight deck. How can first officers be made to intervene in a timely and firm way that will not exacerbate the situation? How may captains be encouraged to act so as to encourage co-pilots to air their uncertainties? Many airline are now tackling these problems with programmes of simulator training that attack not only the rule-based elements of flying (shutting down an engine after a fire alarm), but that are aimed to encourage crews to solve unexpected and sometimes vaguely defined problems on a group basis. Such simulation exercises are frequently video-taped to enable, for example, the captain who believes himself to be exercising an appropriate level of authority to see that he is actually presenting himself as something of a martinet. These exercises also enable the naturally submissive first officer to discover, in a benign role-playing environment, that it is possible to air his views and anxieties in ways that contribute to the effectiveness of the team operation without appearing to challenge the authority of the aircraft commander.

The benefits of improving crew coordination training, and enabling crews to practise solving the types of ill-defined problems that are known to have caused accidents, are difficult to quantify as it is extremely difficult to identify the accident that fails to occur. Nevertheless, such training is becoming widespread and generally accepted in airlines, and is likely to become a legal requirement in airline operations in the foreseeable future.

System factors

The issue of teamwork training described above represents another example of the type of factor that confronts the operator with a decision that requires cost to be balanced against safety benefit. It is relatively easy for the profitable airline to decide that such training is beneficial, but the airline operating in a more competitive area of the aviation system, where economic margins are extremely constrained, may simply be unable to undertake all of the desirable training and standardization of equipment without going out of business. The regulatory authority may have considerable difficulties in compelling such airlines to undertake costly procedures as the airlines may accurately point out that by doing so they will be made less cost efficient vis a vis foreign operators (possibly operating in a less regulated environment) with whom they compete directly.

The temptation for operator and regulator alike, when faced with an acknowledged but intractable problem, is to undertake some unconscious dissonance resolution by regarding the problem as less serious than they might if it were readily soluble. The example of this provided by the operator quoted above shows that this dissonance resolution can extend to the stage of what might even be termed denial. Although CHIRP was not originally designed to counter such behaviour, it has an important role in doing so.

By enabling pilots to report their anxieties to an agency outside the system within which they operate, constraints on candour are removed, and issues that were previously discussed only in crew rooms and bars become available to be fed back into the system in an overt way. However widespread covert knowledge of a problem may be, it is unlikely to generate remedial action since an operator can scarcely be expected to solve a problem with which he has not been overtly acquainted. A good example of this process is provided by the fatique reports submitted to CHIRP. A number of these reports are of incidents in which complete airline crews found themselves asleep while, for example, crossing the Atlantic. Although flying mythology acknowledged such incidents, it has required the CHIRP system to force the problem of the sleeping crew to be confronted and tackled.

Operation of the CHIRP system suggests that the dominant organizational factor of importance to system safety is attitudinal. Management must be forgiven if, when publicly confronted with a safety problem, they seek to minimize its magnitude, but they must not let themselves believe their own publicity. It should be the responsibility of all managements to ensure that when deficiencies in design or operational procedures are reported, they do not seek to discipline, belittle, or even dismiss individuals so as to maintain the status quo, but attempt instead to come to an understanding of the problem and its likely consequences to ensure that any possible modifications and improvements may be made.

Conclusions

This paper has attempted to show that human error on the flight deck requires solution at all levels of the aviation system. The examples provided of failures in equipment design, failures in crew coordination, and failures of safety consciousness in system managers suggest that we should not be considering whether there is a particular psychology of human error, but that we should be attempting to marshal all of the psychological knowledge available to solve the perceptual, skill, design, selection, training, social and organizational problems with which the aviation psychologist is presented.

An important tool for the psychologist involved in studying failures in any complex system must be some form of confidential reporting system for human error. To be effective, such a programme must be operated by an agency external to the system but, if successful, such schemes can not only yield information that enables individual system shortcomings to be tackled, but can compel the whole level of safety consciousness in the system to increase. Industry is now so complex that only by involving psychologists closely in the investigation and analysis of incidents and accidents will they achieve the level of applied knowledge that enables a real and practical contribution to system safety to be made. Aviation has fully adopted this philosophy, and, fortunately, other industries are rapidly following suit.

References

AAIB 1989 (Air Accidents Investigation Branch) Report on the incident involving BAC 1-11 G-AYWB and Boeing 737 EI-BTZ on 12 April 1988 at Gatwick Airport. Aircraft Accident Report 2/89. London: HMSO.

Foushee, H. C. 1984 Diads and triads at 35000 feet: factor affecting group process and aircrew performance. *Am. Psychol.* **39**, 885–893.

Green, R. G. & Farmer, E. W. 1985 Attitude indicators and the ambient visual system. Report of the XVI Conference of the Western European Association for Aviation Psychology, Helsinki.

O'Connor, A. 1987 (January) Safety regulation: an unnecessary burden – or is it? *Airway* (CAA London), 4–6.

Sanderson, P. M. & Harwood, K. 1988 The skills, rules and knowledge classification: a discussion of its emergence and nature. In *Tasks, errors and mental models* (ed. L. P. Goodstein, H. B. Andersen & S. E. Olsen), pp. 21–34. London: Taylor & Francis.

Taylor, R. M. 1988 Trust and awareness in human–electronic crew teamwork. Proceedings of the Conference on 'The human-electronic crew member: can they work together?' Ingolstadt (F.R.G.), September 1988.

Wheale, J. 1983 Crew co-ordination on the flight deck of commercial transport aircraft. In *Flight operations symposium* (ed. N. Johnson). Dublin: Irish Airline Pilots Association/Aer Lingus.

Phil. Trans. R. Soc. Lond. B **327**, 513–518 (1990)
Printed in Great Britain

Medication and skilled work

By A. N. Nicholson
Royal Air Force Institute of Aviation Medicine, Farnborough, Hampshire GU14 6SZ, U.K.

There is increasing interest in the way in which drugs impair performance. This has arisen because some may impair day-to-day skills of those whose occupations demand vigilance and motor skill, and of those who are involved in decision making or where interpersonal relations are crucial. For many years the position was adopted, at least in certain occupations where impaired performance could be a danger to others, that the use of any drug should preclude employment. However, recent advances in therapeutics and a greater understanding of drug action in man has made this rather uncomplicated view of life less tenable, and there is now an increasing desire that advances in therapy should, if at all possible, be available to occupational groups, such as airline pilots. In this way the adverse effect which a drug may have on performance has become an important aspect of its clinical profile.

Hypnotics appropriate for transient insomnia, which may arise from the irregularity of rest inherent in many occupations, need to be free of residual effects, antihistamines that are sedative must be avoided, and drugs used in the management of mild hypertension, often during the important years of middle life, must be as free as possible from central effects. And it must be emphasized that these drugs are often used by active, healthy or near healthy individuals. The issues involved in the safe use of a particular drug by a particular individual are complex, and as with all aspects of therapeutics it is sometimes necessary to balance efficacy and adverse effects.

Introduction

The purpose of this paper is to discuss the issues involved in making the decision of whether a certain drug or drugs can be used safely. Hypnotics and antihistamines present examples of the two main problems which arise. In the case of antihistamines, whether individuals can work safely under the influence of a drug, and in the case of hypnotics, how long after ingestion of a drug that is known to impair performance it is safe to carry out skilled work. The aim is to show the current approach to the safe use of drugs, and to outline the ways that may be adopted to ensure that a drug can be used by those involved in skilled work.

Methodology

Broadly speaking, there are two approaches to the study of performance, and also for predicting the effect that a drug has on the day-to-day life of an individual. The activity of a drug can be built up as a profile by using a variety of laboratory tests directed to assessing specific skills relevant to the work of the individual, or the skill itself may be simulated with as much accuracy as possible. In the two approaches there are common considerations of methodology.

In all such studies dose- and time-response data are needed, and the dose range must be relevant to the projected therapeutic use. In the case of centrally acting drugs it may be appropriate to include a dose which, though outside the anticipated therapeutic range, is high enough to impair performance, as this would give an indication of the margin of safety. Dose- and time-related studies have the advantage that the analysis of the data is more likely to detect a relevant effect than analysis of single assessments, and that variations in performance related to the circadian activity of the individual are taken into account. Measurements must, of course, be related to ingestion of placebo, and an active control should be used to ensure that absence of impaired performance is not due to relative insensitivity of the testing procedures.

A large variety of tests has been used in the assessment of impaired performance. These include the deceptively simple paper and pencil tests, such as digit symbol substitution, tests that assess memory and attention, and those that assess psychomotor skill. Tests are also used which relate to specific senses such as vision. Those which measure a neurological entity such as body sway, eye movement, flicker fusion, electrical activity of the brain, and more recently, drowsiness (sleep latencies during the day), are also included. These avoid the problems of measuring performance, and so may be useful when there are difficulties in measuring performance, as in the elderly. However, more needs to be known about the significance of changes in such functions to the overall capability of the individual before they can be used alone in the study of impaired performance with drugs.

As to the choice of tests there are two main issues: information on the skill impaired and on the persistence of effect (evidence of modified central activity) is needed. If information on persistence of effect is sought, no matter how minimal, accurate and well-designed studies with pencil and paper are often useful. Psychomotor skills related to the peripheral nervous or oculomotor systems are sensitive tests of skill impairment.

Inevitably the question arises concerning the relevance of performance tests carried out in the laboratory to the day-to-day work of the individual. Many laboratory tasks such as reaction time, pursuit rotor, motor coordination and divided attention may have mere face validity (Starmer & Bird 1984). The study of perception and cognition has been specially advocated (Moskowitz 1984) because the majority of driver-related errors fall into the category of information failures, and attention is the most frequently cited area in alcohol-related accidents. At a first look simulation would appear to be a more attractive approach, but there are serious concerns whether simulation is particularly relevant to the real situation. Indeed, it must be questioned whether studies using simulators are as useful as laboratory tests in providing reliable information on drug effects.

Simulation

Simulation of an occupation or of a day-to-day skill may bring increased reality and motivation to the participation of the subject, but it may nevertheless lack sensitivity in the measurement of performance itself. In a recent study, an antihistamine failied to impair driving performance, whereas it impaired performance on an adaptive tracking task (Cohen *et al.* 1984). The sensitivity of a simulation to centrally acting drugs must always be established. An absence of an effect is of limited value unless it is known at what dose of that drug, or a similar drug, performance would be impaired. Similarly, information on efficacy and adverse effects spread over the whole of the therapeutic dose range is needed. Dose- and time-response data, as in all studies with drugs, are essential.

Clearly, uncertain or insensitive measures obtained with simulation have no advantage over accurate measures from the laboratory. It must be realised that simulators (including vehicle handling tests) are often testing isolated functions in a complex, expensive and uncontrolled way. As pointed out (Starmer & Bird 1984), driving includes the skills of visual search and recognition, information processing under variable load, risk taking, decision making and motor control. There are many factors that influence measurement in a complex situation, and these, together with the inherent variability of a situation that involves simulation, can lead to difficulties in establishing a drug effect.

It is important that a spurious confidence in the use of simulation and in vehicle handling tests does not arise. As far as car accidents are concerned, they are only seldom related to loss of control. It may be far more important to study the decision as to whether a specific manoeuvre is thought to be possible than the ability to carry out accurately the manoeuvre itself, and to study the ability to cope with an unexpected situation than the skill involved in negotiating stationary obstacles. The studies of Cohen (1966) and of Brown *et al.* (1969) have shown that the number of times a vehicle may be driven successfully through a gap of any particular size may not be affected by alcohol, but the inebriated driver attempts more gaps smaller than the size of the vehicle. Further, as pointed out by Walsh (1984), though drugs may only impair well-learned behaviour at high doses, they may impair coping or learning skills at low doses, and this is clearly relevant to the ability to cope with mechanical failure or if someone runs in front of a car.

There are other considerations that must also be borne in mind. A simple handling experiment does not represent the complexity of driving experience, different skills are needed for different types of driving, and an immediate transfer of experience between simulated driving and real driving is unlikely to exist. Personality may also have an important part to play in the genesis of accidents. Reaction times are faster in young people but slower in the elderly, yet accident incidence is the other way around. It is also essential that the more subtle effects of drugs on man are borne in mind. Effects on decision making and on the behavioural integrality of man have yet to be adequately explored, and the impairment of such complex skills may be of considerable significance to the individual.

Cohen *et al.* (1984) have pointed out that, though impairment of real or 'off road' driving has been produced by moderate to large doses of benzodiazepines (Zezulka & Wright 1982), barbiturates and major tranquillizers (Betts *et al.* 1972), antihistamines and lower doses of benzodiazepines sufficient to cause sedation in patients and volunteers (Peck *et al.* 1975) have failed to impair driving performance (O'Hanlon *et al.* 1982; Hindmarch 1976). We must, therefore, always question whether failure to demonstrate impaired vehicle handling means that a drug is unlikely to impair driving or cause accidents.

Nevertheless, there is some measure of agreement between careful studies that use simulation and similar studies with laboratory tests. It is the contention that laboratory tests do not provide such useful information as simulation or vehicle handling that causes most dispute. Laboratory tests can measure skills that need to be preserved and which, in the case of accidents, may be identified from epidemiological studies, but with simulation, careful assessment of its relation to the overall task in question is needed. There is a need to integrate data from different methodologies, and it would be useful if a simple test that has been used widely in studies on the central effects of drugs was included in all studies. Information with a relatively simple task known to be sensitive to centrally acting drugs, such as digit symbol substitution, would allow comparative information to be built up between centres that use laboratory tests and simulation. For the moment, laboratory tests are preferred for the analysis of drug effects, as they provide reliable information on the nature of the skill impaired and on persistence of the effect of a drug (Broadbent 1984).

Clinical pharmacology

The approach used to explore the effects of a drug on performance must relate to the way in which it is to be used. It may be relevant to establish the residual effects of a drug, or it may be important to establish whether it is safe to work while the drug is acting. If persistence of activity is the issue (and hypnotics are a useful example as they are used overnight), the question is whether it is safe to carry out skilled work the next day. Duration of action depends on absorption, distribution into the tissues and elimination not only of the parent drug, but also of its metabolites, and consideration of both pharmacokinetic data and pharmacodynamic studies is necessary before recommendations that relate to safe use can be made with confidence. For instance, the pharmacokinetic profile and pharmacodynamic effects are essential to define accurately the likelihood of residual daytime impairment (Nicholson 1986).

Although hypnotics may be used safely as long as their effect does not persist beyond the sleep period, many drugs are used during the period of work itself, and so may easily affect performance. An example is the antihistamines that are used for their peripheral anti-allergic properties, though they often lead to drowsiness and impaired performance (Nicholson 1983). Drowsiness with the H1-antihistamines has been attributed to various mechanisms such as inhibition of histamine *N*-methyltransferase and blockade of central histaminergic receptors, though serotonergic antagonism, anticholinergic activity and blockade of central alpha adrenoceptors may also be involved.

Whatever may be the cause of sedation with antihistamines the central effects are dependent on the ability of a particular drug to cross the blood–brain barrier, and may pass with ease. The solution would be to develop antihistamines that have difficulty in crossing the blood–brain barrier. Indeed, such antihistamines are now available, for example, terfenadine, astemizole and loratadine (Bradley & Nicholson 1987; Nicholson 1982; Nicholson & Stone 1982), and are free of central effects.

Interpretation

When the presence or absence of impaired performance has been established the findings need careful interpretation. Impaired performance not only implies impairment of a particular skill, but also that the central nervous system has been modified. Other skills that are less obvious and less easily measured may also be impaired. In this context, it is timely for the clinical pharmacologist to review with the psychologist the current approach to the measurement of behaviour and performance, and to ensure that current knowledge is being taken into consideration. Inability to show impaired performance does not necessarily mean a drug is free of adverse effects, as there is no test or group of tests that show whether human performance *in toto* is preserved.

It must also be appreciated that the data on impaired performance with drugs may be obtained in young healthy adults. Further, the effects of centrally acting drugs may vary with age and gender and the ability of the individual to metabolize and excrete the drug, and effects may be enhanced by the concomitant use of other drugs. Impairment of the ability to metabolize a drug as in those with renal failure must also be taken into consideration. There is also clear evidence that the elderly are more sensitive to psycho-active drugs, and the effects of drugs may vary from young adulthood to middle age.

There are no simple answers to the questions that are raised by performance studies with drugs. We must be wary of studies which claim absence of performance deficits, and interpret with caution studies that claim to show impairments. It is important that sensitive techniques are used. Ideally, a drug in which impaired performance of any nature cannot be demonstrated in an adequately designed experiment is required. However, other things being equal, limited impairment of performance may have to be accepted, and the drug that is least likely to impair performance may have to be chosen.

General considerations

Though rapidly eliminated hypnotics and antihistamines free of central effects are now used widely by those involved in skilled activity, problems remain with many other drugs. Particular difficulties arise with psychoactive drugs, such as antidepressants and anxiolytics, which are used for their daytime effects. Of course, in patients with disorders of mood, treatment itself may lead to an improvement in performance, but the possibility of impairment with these drugs most certainly arises if they have sedative effects. The emergence of antidepressants free of sedation has been helpful, and some of the newer anxiolytics may have less sedative effects than others.

There is, however, increasing concern that some drugs without obvious central effects, such as the beta-adrenoceptor antagonists, may have central effects of a subtle nature (Currie *et al.* 1988; Nicholson *et al.* 1988). In this context disturbance of memory should be borne in mind. The safe use of these drugs may well rest with compounds that cross the blood–brain barrier with difficulty, though the question also arises whether tolerance to drugs of this group, even if they cross the blood–brain barrier very slowly, develops in the same way as it would appear to do so with other centrally active drugs. Much more needs to be known about drugs now being used in the management of hypertension, and care may have to be taken in their use.

Most certainly a cautious approach to the use of drugs is essential in the proper management of those who are engaged in occupations where impaired performance would not be acceptable. The possibility that the individual may respond adversely with any drug must always be excluded, even for those drugs free of any experimental evidence of a central effect. Nevertheless, treatment has often to be decided on the basis of that which is least likely to cause harm, and it is fortunate that there are often many drugs available which, though they may have similar therapeutic efficacy, have different effects on performance.

References

Betts, T. A., Clayton, A. B. & Maclay, G. M. 1972 Effects of four commonly used tranquillizers on low-speed driving performance tests. *Br. med. J.* **4**, 580–584.

Bradley, C. M. & Nicholson, A. N. 1987 Studies on the central effects of the H1-antagonist loratidine. *Eur. J. clin. Pharmacol.* **32**, 419–421.

Broadbent, D. E. 1984 Performance and its measurement. *Br. J. clin. Pharmacol.* **18**, 5S–9S.

Brown, I. D., Tickner, A. H. & Simmonds, D. C. V. 1969 Interference between concurrent tasks of driving and telephoning. *J. appl. Psychol.* **53**, 419–424.

Cohen, A. F., Posner, J., Ashby, L., Smith, R. & Peck, A. W. 1984 A comparison of methods for assessing the sedative effects of diphenhydramine on skills related to car driving. *Eur. J. clin. Pharmacol.* **27**, 477–482.

Cohen, J. M. 1966 *A new introduction to psychology*. London: Allen and Unwin.

Currie, D., Lewis, R. V., McDevitt, D. G., Nicholson, A. N. & Wright, N. A. 1988 Central effects of beta-adrenoceptor antagonists. I: performance and subjective assessments of mood. *Br. J. clin. Pharmacol.* **26**, 121–128.

Hindmarch, I. 1976 The effects of the sub-chronic administration of an antihistamine, clemastine, on tests of car driving ability and psychomotor performance. *Curr. med. Res. Opin.* **4**, 197–206.

Moskowitz, H. 1984 Attention tasks as skill performance measures of drug effects. *Br. J. clin. Pharmacol.* **18**, 51S–61S.

Nicholson, A. N. 1986 Hypnotics: their place in therapeutics. *Drugs* **31**, 164–176.

Nicholson, A. N. 1983 Antihistamines and sedation. *Lancet* ii, 211–212.

Nicholson, A. N. 1982 Antihistaminic activity and central effects of terfenadine. A review of European Studies. *Arzneimittel-Forsch.* **32**, 1191–1193.

Nicholson, A. N. & Stone, B. M. 1982 Performance studies with the H1-receptor antagonists, astemizole and terfenadine. *Br. J. clin. Pharmacol.* **13**, 199–202.

Nicholson, A. N., Wright, N. A., Zeitlen, M. B., Currie, D. & McDevitt, D. G. 1988 Central effects of beta-adrenoceptor antagonists II: electroencephalogram and body sway. *Br. J. clin. Pharmacol.* **26**, 129–141.

O'Hanlon, J. F., Haak, T. W., Blaauw, G. J. & Riemersma, J. B. J. 1982 Diazepam impairs lateral position control in highway driving. *Science, Wash.* **217**, 79–81.

Peck, A. W., Fowle, A. S. E. & Bye, C. 1975 A comparison of triprolidine and clemastine on histamine antagonism and performance tests in man: implications for the mechanism of drug induced drowsiness. *Eur. J. clin. Pharmacol.* **8**, 455–463.

Starmer, G. A. & Bird, K. D. 1984 Investigating drug/ethanol interactions. *Br. J. clin. Pharmacol.* **18**, 27S–35S.

Walsh, J. M. 1984 Impaired motor activity. *Br. J. Clin. Pharmacol.* **18**, 97S–98S.

Zezulka, A. & Wright, N. 1982 Effects of two hypnotics on actual driving performance next morning. *Br. med. J.* i, 285–852.

Phil. Trans. R. Soc. Lond. B **327**, 519–528 (1990)
Printed in Great Britain

Respiratory virus infections and performance

By A. P. Smith
Laboratory of Experimental Psychology, University of Sussex, Brighton BN1 9QG, U.K.

Minor illnesses, such as colds and influenza, are frequent, widespread and a major cause of absenteeism from work and education. Yet the clinical symptoms of such illnesses may not be so great as to stop people from working or from carrying out everyday activities. It is therefore important to determine whether these viral infections alter central nervous system function and change performance efficiency. Data on this topic are almost non-existent, which in part reflects the difficulties inherent in carrying out such studies. In real life it is almost impossible to predict when such illnesses will occur, and difficult to establish which virus produced the illness.

This problem was overcome by studying experimentally induced infections and illnesses at the Medical Research Centre (MRC) Common Cold Unit in Salisbury. Results from this research programme show that:

(i) colds and influenza have selective effects on performance;
(ii) even sub-clinical infections can produce performance impairments;
(iii) performance may be impaired during the incubation period of the illness;
(iv) performance impairments may still be observed after the clinical symptoms have gone.

These results have strong implications for occupational safety and efficiency and it is now essential to assess the impact of naturally occurring colds and influenza on real-life activities.

1. Introduction

It is widely accepted that the physiological state of the person is very important in determining performance efficiency. This has usually been studied by manipulating some feature of the environment, such as the noise level, or by examining endogenous variations in alertness, or by changing the state of the person by administering drugs. Such studies have usually been carried out with normal, healthy subjects. However, a person's state is frequently changed by infection and illness and it is important to determine whether performance is altered by these. Many illnesses are so severe that the individual is unable to carry out normal activities, and the question of impaired performance does not arise (although the absence of an individual may pose additional problems for other workers). Other illnesses, such as the common cold and influenza, may not be so severe that they prevent a person from working or carrying out other activities. It is important, therefore, to assess the effects that these respiratory virus infections have on performance efficiency.

Respiratory virus illnesses are a major health problem in that the acute infections and their consequences account for a substantial proportion of all consultations in general practice. These illnesses are also one of the main causes of absence from work and education and we spend over one hundred million pounds a year on medication for them. Colds and influenza affect most of the population (it has been estimated that we have between one and three colds a year) and it is highly desirable to increase our knowledge of these illnesses, with the practical aims of lowering health care costs, decreasing absenteeism from work and improving the quality of life.

Despite the frequency with which such illnesses occur there has been no research on their behavioural effects. This is true of infectious diseases in general and Warm & Alluisi (1967) concluded that 'Data concerning the effects of infection on human performance are essentially non-existent'. Since that review the effects of certain infectious diseases have been studied (Alluisi *et al.* 1971, 1973; Thurmond *et al.* 1971). In one experiment (Alluisi *et al.* 1971) volunteers were infected with *Pasteurella tularensis* (which produces rabbit fever, a febrile disease characterized by headache, photophobia, nausea, myalgia and depression). Those who became ill showed an average drop in performance of about 25% and after recovery they were still 15% below the level of the control group. There was also evidence that some types of activity were more impaired than others, with active tasks such as arithmetic computation showing a greater decrement than passive tasks such as watchkeeping. The illnesses studied in these experiments are very severe and analogous effects must rarely occur in everyday life. In contrast to this, colds and influenza are frequent and widespread and yet we have little information on whether they reduce the efficiency of performance.

One possible reason for this lack of research is that people feel they already know about the behavioural effects of such illnesses and it is, therefore, a waste of time carrying out experiments on this topic. One commonly held view is that when someone is ill everything is performed less efficiently than normal. An opposite view is based on the observation that people often go to work when suffering from these illnesses, which suggests that any effects are likely to be too slight or transitory to be of practical importance. An alternative view emerges from the literature on stress and performance. Recent research on the effects of changes in state has shown that different activation states produce distinct profiles of performance changes (see Hockey & Hamilton 1983). Colds and influenza are different in that colds produce local symptoms (nasal secretion) whereas influenza also gives rise to systemic effects. One might, therefore, predict that the two types of illness will change performance in different ways.

Another reason why there has been little research in this area is that it is difficult to study naturally occurring illnesses. These illnesses are hard to predict and it is often unclear which virus is the infecting agent (there are over 200 viruses that produce colds). The study of such illnesses only enables one to examine the effects of clinical illnesses, and it is often difficult to obtain objective measures of the symptoms. It is also possible that sub-clinical infections may influence behaviour and these can only be identified by using the appropriate virological techniques. These problems were overcome here by examining the effects of experimentally induced colds and influenza at the Medical Research Centre (MRC) Common Cold Unit, Salisbury. The main features of the methodology and routine of the Common Cold Unit are summarized in the following section.

2. The routine of the common cold unit

The volunteers, who are aged 18–50 years, come to the unit for a 10-day stay, during which time they agree to receive an infecting virus inoculum. The volunteers are housed in groups of two or three and isolated from outside contacts. Before the visit the volunteers supply the unit with a self-reported medical history. People taking sleeping pills, tranquillisers or anti-depressant medicines are not allowed to take part in the trial, and neither are pregnant women.

On the first day of the visit they undergo a medical examination, and any who fail this are excluded from the trial. A blood sample is also taken at this time to enable assessment of initial antibody level. Isolation begins in the afternoon of the first day and the volunteers are observed during a three-day quarantine period so that any individuals who are incubating a cold may be excluded. A nasal washing is obtained on the third day of the trial and individuals with sub-clinical infections are also excluded.

Volunteers are usually given the virus or saline placebo on the fourth day. The trials are conducted double-blind with neither the volunteers, the unit's clinician nor any of the personnel who interact with the subjects knowing which volunteers received virus or placebo. Volunteers are usually administered 10–100 TCD50 of a rhinovirus by nasal drops. The most common serotypes used are RV9 and RV14, and sometimes both are given. Coronaviruses and respiratory syncytial viruses are used in other trials. Influenza trials are carried out less frequently, but there have been studies of influenza A and influenza B viruses.

Following virus challenge there is an incubation period of 24–96 h depending on the type of virus. In general, about one third of the volunteers develop significant symptoms and one-third have sub-clinical infections. Very few volunteers are given placebo because one third of those given the virus remain uninfected.

On each day of the trial the severity of cold or influenza symptoms is assessed by the unit's clinician. Self-reported respiratory symptoms are also collected on a standardized paper and pencil instrument. Objective measures of symptomatology are also recorded, namely the number of paper handkerchiefs used, the weight of nasal secretion, and sub-lingual temperatures. At the end of the trial the clinician decides whether volunteers have had significant colds or not (according to well-established procedures, see Beare & Reed (1977)). Nasal washings are taken so that the shedding of the virus can be assessed, and a blood sample is returned to the unit three weeks after the visit to allow the antibody level to be measured again.

All procedures of the unit are approved by the Harrow District Ethical Committee and carried out with the true consent of the volunteers.

3. Performance testing

The data reported in this article were collected during clinical trials that had other specific aims. This imposed certain limitations on the type of task and frequency of testing. Similarly, portable tests had to be used because the volunteers were in isolation and could only be tested in their flats.

Two main methods have been used. The first involved administration of paper and pencil tests measuring logical reasoning, visual search and semantic processing at four times of day on every day of the trial. It was important to examine performance at several times of day for two reasons. First, it has been shown that performance changes over the day (see Colquhoun 1971), and secondly, there is diurnal variation in the severity of symptoms of colds and influenza (see Smith *et al.* 1988*a*), with nasal secretion and temperature being greatest in the early morning.

The second method of assessing performance used computerized performance tasks, with volunteers being tested once in the pre-challenge quarantine period and once when symptoms were apparent in some volunteers. In certain trials performance testing was also carried out in the incubation period. Subjects were always tested at the same time of day on all occasions (although some subjects were tested in the morning and others in the afternoon, which meant that diurnal variation could be examined). The computerized tasks were selected to assess a range of functions (memory, attention and motor skills) and most of the tests have been widely used in studies of stressors or drugs.

Analyses of covariance have been carried out on the data using the pre-challenge scores as covariates. This statistical technique takes account of baseline differences when assessing the effects of the illness.

4. Early results

Smith *et al.* (1987*a*) compared the effects of colds and influenza on different aspects of psychomotor performance. Two of the tests required subjects to detect and respond quickly to targets appearing at irregular intervals (a variable fore-period simple reaction time task and a 'fives' detection task). The other was a pursuit tracking task designed to test hand–eye coordination.

Influenza B illnesses increased reaction times in both detection tasks. These results are shown in table 1. Analysis of the tracking task showed no significant difference between those with influenza and those who remained healthy.

Table 1. Mean reaction time (milliseconds) in the simple reaction time and fives detection tasks for the influenza and uninfected groups

	influenza	uninfected
simple reaction time task		
pre-challenge	320	348
post-challenge	503	325
fives detection task		
pre-challenge	460	421
post-challenge	573	409

The effects of colds on the three tasks are shown in table 2. The difference between the volunteers with colds and the uninfected subjects was not significant for either detection task. However, volunteers with colds were much worse at the tracking task than those who remained well.

Table 2. Mean scores for volunteers with colds and uninfected subjects on the three performance tasks

(Figures given for simple and fives tasks are in milliseconds.)

	colds	uninfected
simple reaction time task		
pre-challenge	332	329
post-challenge	326	308
fives detection task		
pre-challenge	399	402
post-challenge	392	398
tracking task (number of contacts)		
pre-challenge	13.8	13.3
post-challenge	12.6	21.5

These results show that there are effects of respiratory virus illnesses on performance, but that the functions affected depend on the nature of the illness. Many types of behaviour involve both hand–eye coordination and the ability to quickly detect unexpected stimuli. Such tasks may be susceptible to the effects of both colds and influenza. It should be pointed out that the effects of these illnesses were very large. For example, influenza led to a 57% impairment on the reaction time task and a moderate dose of alcohol or having to perform at night would typically produce a 5–10% impairment. The results are of theoretical interest because of the dissociation of functions by the different illnesses. Possible mechanisms underlying these effects are discussed in a later section.

Smith *et al.* (1988*b*) compared the effects of colds and influenza on performance of the paper and pencil tests at four times of day. The main results may be briefly summarized as follows. Influenza B illnesses impaired performance on a visual search task with a high-memory load (subjects had to search for the presence of any of five target letters at the start of a line and the speed and accuracy of searching a 12-line block was recorded). This effect is shown in table 3.

TABLE 3. MEAN TIMES (SECONDS) TO COMPLETE THE PEGBOARD AND SEARCH AND MEMORY TASKS FOR VOLUNTEERS CHALLENGED WITH AN INFLUENZA B VIRUS

(Scores are the adjusted means from analyses of covariance.)

	pegboard	search and memory
uninfected	39.4 s	94.8
sub-clinical infection	45.0	108.3
influenza	42.6	135.8

In contrast to this, volunteers with influenza were not impaired on a test of manual dexterity and movement time (the subject had to transfer pegs as quickly as possible from one pocket solitaire set to another). Volunteers who developed colds after challenge with respiratory syncytial virus were impaired on the pegboard task but not on the search task. The pegboard data are shown in figure 1. Neither influenza nor colds impaired the speed or accuracy of logical reasoning nor semantic processing.

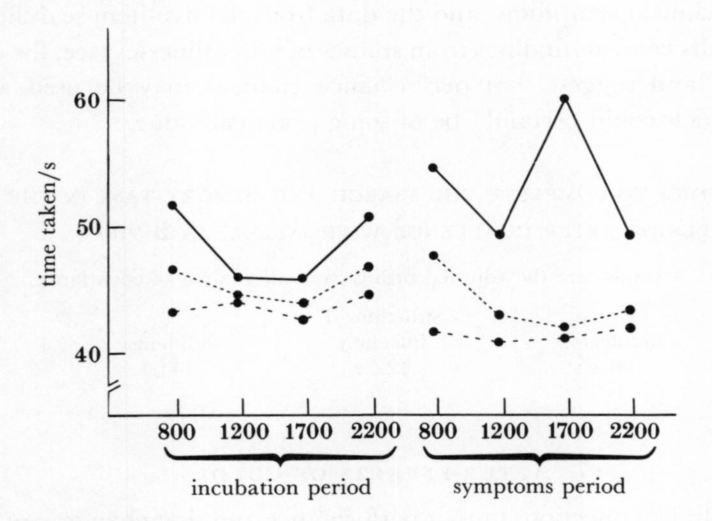

FIGURE 1. Effects of colds on performance of the pegboard task at four times of day during the incubation and symptoms period. (——), clinical group; (----), sub-clinical group; (–––) uninfected group. (Times are the adjusted means from analyses of covariance.)

5. Sub-clinical infections

One of the great advantages of studying experimentally induced respiratory infections is that it is possible to examine the effects of sub-clinical infections. Table 3 shows that those volunteers with sub-clinical influenza B infections were slower on the five-item search task than those who remained uninfected. This effect of sub-clinical infection was not observed in the colds–pegboard data. However, results from another experiment suggest that sub-clinical infections with cold-producing viruses can also impair performance. This experiment (Smith *et al.* 1987 *b*) examined the effects of colds and sub-clinical infections on performance of the five-choice serial response task. The subject was shown five boxes in a row across the computer screen and when a black square appeared in a box the subject had to press the corresponding key on the computer keyboard. As soon as the subject responded the square reappeared and the subject had to make the next response. The results are shown in table 4, and it can be seen that both those volunteers with significant colds and those with sub-clinical infections were slower than the uninfected group.

Table 4. Mean number of responses per minute in the five-choice serial reaction time task on symptomatic days

(Scores are the adjusted means from analyses of covariance.)

	colds	sub-clinical infection	uninfected
symptomatic days	78.4	78.8	81.6

6. Performance changes during the incubation period

Another advantage of doing this research at the Common Cold Unit is that one knows when the virus was given and one can, therefore, examine changes during the incubation period. Results from the incubation period of the influenza B study showed that behavioural changes precede the onset of clinical symptoms, and the data from the five-item search task are shown in table 5. These results confirm findings from studies of other illnesses (see, for example, Elsass & Henriksen 1984), and suggest that performance changes may be used as indicators of subsequent illness which could certainly be of some practical value.

Table 5. Mean times to complete the search and memory task in the incubation period after challenge with influenza B virus

(Times (seconds) are the adjusted means from an analysis of covariance.)

uninfected	sub-clinical infection	influenza
106.4 s	122.2	144.4

7. After-effects of colds

The fact that sub-clinical infections impair performance and that changes are observed in the incubation period shows that viral infections may alter performance in the absence of clinical symptoms. This issue was also examined by measuring the after-effects of colds on performance

(Smith *et al.* 1990*a*). In this trial volunteers stayed at the unit for three weeks and it was possible to test them not only when they were symptomatic but also one week after the symptoms had gone. The volunteers carried out choice reaction time tasks and those with colds were slower than those without symptoms (see table 6), and this difference was still present one week later, even though the clinical symptoms had gone (see table 6).

Table 6. Effects and after-effects of colds on choice reaction time (milliseconds) performance

	colds	no colds
day 2, pre-challenge	511	523
day 7, symptoms present	530	500
day 14, symptoms no longer present	511	487

These results confirm the findings of Alluisi *et al.* (1971) in showing that effects of viral illnesses continue into convalescence. Grant (1972) has also reported similar after-effects of naturally occurring influenza illnesses. At the moment it is unclear why one gets after-effects of these illnesses. One possibility is that the performance tests are sensitive to the immunological changes that occur after the symptoms have gone. Another possibility is that it represents a carry-over effect, with subjects continuing to perform in the same way as when they were symptomatic. Further experiments are required to resolve this issue.

8. Mechanisms underlying the effects of respiratory virus infections on performance

In influenza, interferon alpha can be found in the circulation and it is now clear that such peptide mediators have effects on the CNS. It was, therefore, postulated that the changes in performance observed in volunteers with influenza may be due to the effect of interferon on the CNS. This was tested by injecting volunteers with different doses of interferon alpha and it was predicted that those who received a dose which produced influenza-like symptoms would show comparable performance impairments. The results are described in detail in Smith *et al.* (1988*c*) and they showed that an interferon injection of 1.5 Mu produced an identical profile of performance changes to those seen in influenza. The smaller doses produced no significant clinical symptoms. Reaction times on the variable fore-period simple reaction time task were slower for the group with clinical symptoms, whereas the other groups improved over the day. This is shown in figure 2. However, neither the tracking task nor logical reasoning task showed any impairment, which confirms negative results obtained in influenza trials.

While interferon-induced changes in CNS function provide a plausible explanation for the selective effects of influenza on performance, it is less clear which mechanisms underlie the effects of colds. One possibility is that some other lymphokine or cytokine, such as interleukin 1, is involved. Indeed, it has been shown that this mediator has an effect on the muscles which could account for the impaired hand–eye coordination. Alternatively, the impairments could reflect changes in sensory stimulation via the trigeminal nerves in the nose.

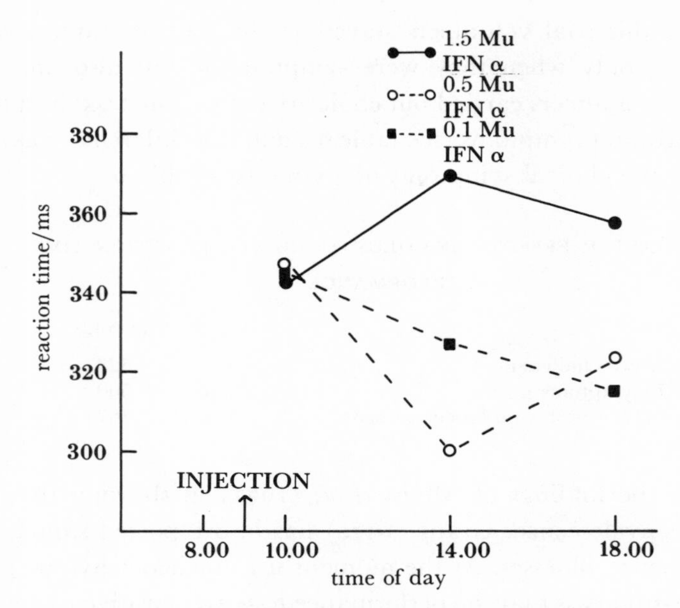

FIGURE 2. Effects of three doses of α-interferon on performance of the variable fore-period reaction time task. (Times are the adjusted means from an analysis of covariance.)

9. RECENT RESULTS

(a) Colds and memory

Most of the early studies used simple psychomotor tasks. In recent trials we have examined the effects of colds on different aspects of memory. Many aspects of memory, such as the ability to recall a list of words, or to recall a string of digits in order, appear to be unaffected by having a cold. Colds do appear to impair the learning and recall of more complex material such as stories, and further studies examining the effects of colds on tasks requiring sustained attention are needed. However, colds do not interfere with recall of material learned prior to infection, and there is some evidence that reminiscence, the ability to subsequently recall information which was not retrievable immediately after learning, is better when the person has a cold. This may reflect the decreased arousal produced by the cold, and other conditions that further reduce arousal, such as the consumption of a large lunch, amplify this effect.

(b) Prediction of susceptibility to illness

Previous research (see, for example, Totman *et al.* 1980; Broadbent *et al.* 1984) has shown that psychological factors are important in determining susceptibility to infection and illness. For example, introverts appear to be more likely to become infected than extraverts. Our research has shown that measures of performance taken before virus challenge are often related to the probability of developing a cold (see Smith *et al.* 1990*b*). In a very recent study we have examined the relation between visual sensitivity and susceptibility to colds. The volunteers were shown a visually stressful pattern (see Nulty *et al.* 1987) and asked to note the illusions it induced. The results showed that subjects who subsequently developed a cold were more sensitive than those who remained symptom free (see table 7). Furthermore, this test is not related to personality traits such as extraversion or neuroticism, and we have, therefore, a relation between the state of the individual and susceptibility to colds.

Table 7. Mean number of pre-challenge visual illusions reported by volunteers who subsequently developed colds and those who remained symptom free

significant colds	doubtful colds	no colds
3.2	2.5	1.5

(c) *Drugs, colds and performance*

Many trials at the Common Cold Unit are designed to assess prophylactic and therapeutic drugs. It is possible to assess not only whether the drug modifies the clinical symptoms but also if it removes the performance impairments associated with a cold. One study has examined whether sodium nedocromil (a drug thought to suppress mediators such as histamine) reduces cold symptoms and influences the extent of the cold-induced performance impairment (see Barrow *et al.* 1990). The results showed that volunteers taking nedocromil had less severe colds and that the drug also reduced the size of the cold-induced performance impairment. Unfortunately, the mode of action of nedocromil is unclear and this study provides little further information on the mechanisms underlying the effects of colds on performance.

10. Conclusions

The research described here has demonstrated that experimentally induced upper respiratory virus infections and illnesses can reduce performance efficiency. The exact effect depends on the activity being performed and the type of virus, with colds producing different effects from influenza. The performance impairments are not confined to volunteers with clinical symptoms. Sub-clinical infections can also reduce performance, although once again these effects are selective. Performance also changes in the incubation period before the emergence of the symptoms and the impairments may persist after the symptoms have gone. We have also found that performance measures taken before virus challenge are related to the susceptibility of developing a cold.

While these studies of experimentally induced colds and influenza show that performance is impaired, we have little information on the role played by naturally occurring infections and illnesses in human error. Such illnesses are typically more severe than those we have examined at the Common Cold Unit and one could argue that they should produce far greater effects. However, many workers are well-practised at their jobs and one could suggest that this will make them less susceptible to the effects of these illnesses. The worker is often exposed to a range of factors which may alter performance efficiency and it is possible that viral illnesses may not only have direct effects on performance, but may indirectly influence it by making the person more susceptible to other factors.

Overall, one may conclude that we now have plenty important information about the effects of upper respiratory virus infections on performance efficiency. These results have strong implications for occupational safety and efficiency, and show that it is now essential to assess the real impact of naturally occurring colds and influenza on real-life activities.

I express my gratitude to all the volunteers who have taken part in these experiments, to the Director and Staff of the Common Cold Unit for their help in the clinical trials, and to Kieran

Coyle, Susan Leekam and Len Armer, who have played major roles in the data collection and analysis of experiments.

References

Alluisi, E. A., Thurmond, J. B. & Coates, G. D. 1971 Behavioral effects of infectious disease: respiratory *Pasteurella tularensis* in man. *Percept. Motor Skills* **32**, 647–688.

Alluisi, E. A., Beisel, W. R., Bartelloni, P. J. & Coates, G. D. 1973 Behavioral effects of tularemia and sandfly fever in man. *J. infect. Dis.* **128**, 710–717.

Barrow, G. I., Higgins, P. G., Al-Nakib, W., Smith, A. P., Wenham, R. B. M. & Tyrrell, D. A. J. 1990 The effect of intranasal nedocromil sodium on viral upper respiratory tract infections in human volunteers. *Clin. Allergy.* (In the press.)

Beare, A. S. & Reed, S. E. 1977 The study of antiviral compounds in volunteers. In *Chemoprophylaxis and virus infections* (ed. J. S. Oxford), vol. 2, pp. 27–55. Cleveland: CRC Press.

Broadbent, D. E., Broadbent, M. H. P., Phillpotts, R. & Wallace, J. 1984 Some further studies on the prediction of experimental colds in volunteers by psychological factors. *J. psychosom. Res.* **28**, 511–523.

Colquhoun, W. P. 1971 Circadian variation in mental efficiency. In *Biological rhythms and human performance* (ed. W. P. Colquhoun), pp. 39–107. London: Academic Press.

Elsass, P. & Henriksen, L. 1984 Acute cerebral dysfunctions after open heart surgery. A reaction time study. *Scand. J. thorac. Surg.* **18**, 161–165.

Grant, J. 1972 Post-influenzal judgement deflection among scientific personnel. *Asian J. Med.* **8**, 535–539.

Hockey, R. & Hamilton, P. 1983 The cognitive patterning of stress states. In *Stress and fatigue in human performance* (ed. G. R. J. Hockey). Chichester: Wiley.

Nulty, D. D., Wilkins, A. J. & Williams, J. M. G. 1987 Mood, pattern sensitivity and headache: a longitudinal study. *Psychol. Med.* **17**, 705–713.

Smith, A. P., Tyrrell, D. A. J., Coyle, K. B. & Willman, J. S. 1987*a* Selective effects of minor illnesses on human performance. *Br. J. Psychol.* **78**, 183–188.

Smith, A. P., Tyrrell, D. A. J., Al-Nakib, W., Coyle, K. B., Donovan, C. B., Higgins, P. G. & Willman, J. S. 1987*b* Effects of experimentally-induced respiratory virus infections on psychomotor performance. *Neuropsychobiology* **18**, 144–148.

Smith, A. P., Tyrrell, D. A. J., Coyle, K. B., Higgins, P. G. & Willman, J. S. 1988*a* Diurnal variation in the symptoms of colds and influenza. *Chronobiol. Int.* **5**, 411–416.

Smith, A. P., Tyrrell, D. A. J., Al-Nakib, W., Coyle, K. B., Donovan, C. B., Higgins, P. G. & Willman, J. S. 1988*b* The effects of experimentally-induced respiratory virus infections on performance. *Psychol. Med.* **18**, 65–71.

Smith, A. P., Tyrrell, D. A. J., Coyle, K. B. & Higgins, P. G. 1988*c* Effects of interferon alpha on performance in man: a preliminary report. *Psychopharmacology*, **96**, 414–416.

Smith, A. P., Tyrrell, D. A. J., Al-Nakib, W., Barrow, G. I., Higgins, P. G., Leekam, S. & Trickett, S. 1990*a* Effects and after-effects of the common cold and influenza on human performance. *Neuropsychobiology*. (In the press.)

Smith, A. P., Tyrrell, D. A. J., Coyle, K. B., Higgins, P. G. & Willman, J. S. 1990*b* Individual differences in susceptibility to infection and illness following respiratory virus challenge. *Psychol. Heal.* (In the press.)

Thurmond, J. B., Alluisi, E. A. & Coates, G. D. 1971 An extended study of the behavioral effects of respiratory *Pasteurella tularensis* in man. *Percept. Motor Skills* **33**, 439–454.

Totman, R., Kiff, J., Reed, S. E. & Craig, J. W. 1980 Predicting experimental colds in volunteers from different measures of recent life stress. *J. Psychosom. Res.* **24**, 155–163.

Warm, J. S. & Alluisi, E. A. 1967 Behavioral reactions to infection: review of the psychological literature. *Percept. Motor Skills* **24**, 755–783.

Phil. Trans. R. Soc. Lond. B **327**, 529–541 (1990)
Printed in Great Britain

Sustained performance and some effects on the design and operation of complex systems

By M. F. Allnutt, D. R. Haslam, M. H. Rejman and S. Green
Army Personnel Research Establishment, Ministry of Defence, Farnborough, Hants GU14 6TD, U.K.

Man, increasingly the limiting element in the military man–machine system, must often operate for several days in a high-risk environment with little or no sleep. It is necessary, therefore, to have some knowledge of the likely effects of sleep deprivation to predict his behaviour and minimize the adverse effects of sleep loss. The early work of the Army Personnel Research Establishment (APRE) concentrated on studying the infantryman in field trials, characterized by more realism and of greater length than previously attempted. Although measures of cognitive functioning were included in these trials, continuous cognitive performance was not assessed, nor was performance on complex tasks. An opportunity to remedy this situation arose because of a newer study concerned with controlling a removely-piloted air vehicle from a ground control station (GCS). A 65-hour experiment was designed during which subjects performed continuously either on the GCS simulator or on a battery of cognitive tests, mood scales, and physiological assessments. Results showed that whereas performance showed the usual deterioration in the test battery, it held up remarkably well on the simulator. Several reasons for this difference are suggested.

Introduction

Man, increasingly the limiting element in the military man-machine system, must often operate for several days in a high-risk environment with little or no sleep. To make optimum use of this limiting resource we must be able to make a fairly accurate prediction about his likely behaviour, so allowing procedures and equipment to be designed to minimize the negative effects of sleep loss. When neither extrapolation from laboratory experiment nor collated observations from real-world situations are sufficient to answer specific questions, we must resort to field trials or simulation. A series of such trials, started in the 1970s, and also a full simulation, form the substance of this paper.

Our work has not been an attempt to confirm or deny any theoretical prediction about sleep loss or stress (see Hartley *et al.* 1989), but by field trial, simulation, extrapolation from other researchers' experiments and evidence from real-life military and analogous situations to provide best evidence as to how soldiers are likely to behave in specific operational situations. Any such work in this area must of necessity lack two critical elements, perhaps best described as fear and the chaos of war, and there is evidence both from real-life observation (Marshall 1947; Bourne 1969) and experiment (Berkun *et al.* 1962; Baddeley 1972) that these elements have a profound effect on behaviour. We can make some general predictions about the likely direction of such effects, for example, a narrowing of focus that might lead to improved performance, irrelevant behaviour, or escape, but the magnitude of such effects is likely to be situation specific.

The field trials

About 14 years ago, the Army Personnel Research Establishment (APRE) was asked a very specific question, 'How much sleep do soldiers require to remain militarily effective for up to 9 days?' The literature, which had been recently reviewed by Johnson & Naitoh (1974), provided a wealth of laboratory data almost exclusively devoted to experiments of less than 48 h duration. Examples of slightly longer duration were provided by Murray & Lubin (1958), Wilkinson (1962, 1964) and Williams & Lubin (1967). The few field trials there were (Banks *et al.* 1970; Haggard 1970; Dudley *et al.* 1974) had been of relatively short duration (less than 5 days) and only two had aimed at military realism (Drucker *et al.* 1969; Ainsworth & Bishop 1971). Historical evidence from battle and expert opinion gave widely differing answers as to how much sleep was necessary to remain militarily effective for nine days, and so we embarked on a series of five trials, two in the field, two in the laboratory and one combined, to provide best estimates of likely performance under agreed military scenarios. Our approach throughout was to use experienced infantrymen as subjects and expose them to a multidisciplinary battery of measures that included military performance, military judgement, cognitive, subjective and physiological measures. The data are necessarily full and complex and are to be found in Haslam (1981, 1982, 1983, 1985*a*, *b*); the aim here is to provide an overview of the earlier work as a lead in to more recent work.

Taking into consideration the results from earlier studies, the conditions for the first field trial were 0, 1.5 and 3 h sleep per 24 h. All the 0-h sleep platoon were judged to be militarily ineffective after 48 h without sleep and had withdrawn from the trial after 100 h, while 50% of the 1.5-h platoon survived the 9 days but with major degradation of performance; 90% of the 3-h platoon survived, and with less deterioriation in performance. Figure 1 shows a typical result, this being for a vigilance shooting task that required sustained attention for 20 min. In this first trial we allowed unlimited recovery sleep while in a subsequent one we studied the

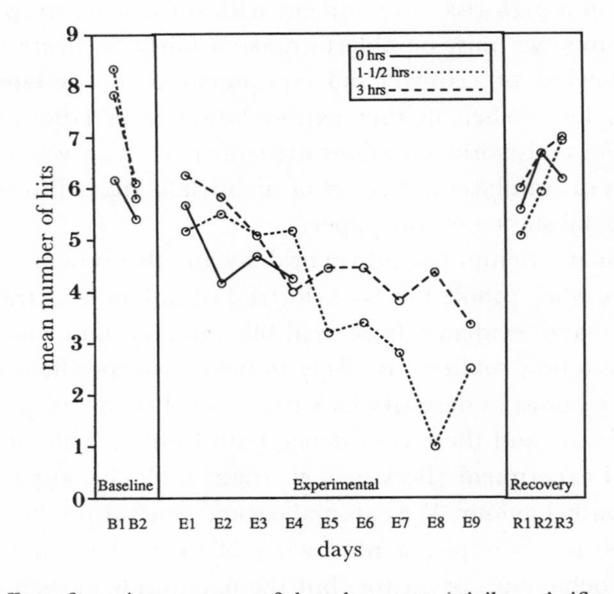

FIGURE 1. The effect of varying amounts of sleep loss upon 'vigilance' rifle-shooting performance, mean number of hits for the three platoons.

more likely regime of total followed by partial sleep loss (Haslam 1982). Four hours sleep after 90 sleepless hours was very restorative, while three 4-h sleep periods over 72 h after 90 h without sleep restored performance on a battery of tasks to 88 % of its baseline value. Figure 2 shows a typical result, this one being for a military de-coding task. In a further trial another realistic regime was followed, namely, partial followed by total sleep loss (Haslam 1985 *b*).

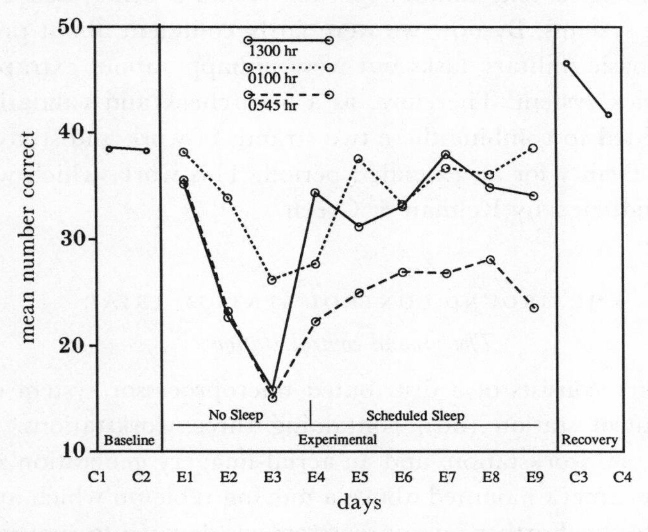

FIGURE 2. The effect of sleep loss and 'recovery' sleep upon decoding grid references, mean number correct for 10 subjects at different times of day.

A question often asked by military commanders is whether several short periods of sleep are as beneficial as one longer uninterrupted one. The results from a 5-day laboratory trial (Haslam 1985 *a*) showed that there were no significant differences in cognitive and mood scores between two groups, one of whom was scheduled 4 one-hour naps, and the other one 4 h continuous sleep per 24 h. Further, neither group showed a significant difference from baseline values, showing the value of 4 h sleep either in one uninterrupted block or in 4 scheduled one-hour naps per 24 h.

From this series of trials and incorporating data from sister establishments in the U.S.A. and Canada, APRE has been able to produce two simple, pocket-sized guides for the military, one on the likely effects of sleep loss and the other on sleep management. These have been issued widely throughout the Army.

The link to more complex tasks

The recent rapid proliferation of high technology equipment on the battlefield now means that nearly all soldiers have to operate complex systems. A common and most critical example of this is the small team who must collate and act upon a wealth of data generated by sophisticated sensor systems. These personnel must maintain a constant high level of performance, be flexible in their approach, and be able to operate in isolation for extended periods.

In parallel with our work on sustained operations, APRE has had a long-standing commitment to design a ground control station to acquire and process data gathered from a remotely piloted air vehicle system. A comprehensive task analysis, attention to ergonomics, and experimentation on the man–computer interactions produced a prototype system that has been steadily improved over the years by being operated in realistic laboratory simulations by teams of soldiers who will operate the real system when it enters service in two years' time (Rejman & Ford 1986). Thus the current ground control station is the product of an iterative design process by psychologists and military specialists, and is being used as the generic basis for several other future systems. By now we were fairly confident about predicting the main effects of sleep loss on basic military tasks but were unhappy about extrapolating to a team operating such a complex system. Therefore, as a final check and validation of our ground control station, we decided to combine these two strands of work and study the performance of teams operating the facility for an extended period. This work, which will now be briefly described, has been conducted by Rejman & Green.

The ground control station trial

The ground control station

The simulation facility consists of a distributed microprocessor system containing a fully interactive ground control station (GCS) containing three workstations, a higher level of command facility with one workstation, and an aerial-imagery generation system. This latter sub-system consists of a camera mounted above a moving table on which are displayed aerial reconnaissance photographs. Further microprocessors are devoted to running a mathematical model of the air vehicle characteristics and handling the communications traffic. Tasks were in the form of areas to be overflown and reported upon. Thus a representative GCS crew can plan and execute target acquisition and intelligence-gathering missions realistically, including complex tasks such as mission allocation, based on temporal and geographical considerations, photographic interpretation, map reading, navigation, and air vehicle control.

Subjects

Five 3-man crews were used as subjects. They consisted of experienced non-commissioned officers from the projected user population. While highly skilled in many aspects of the task, they were entirely new to this simulation.

Procedure

The Army wished us to investigate a 3 day/2 night scenario of 65 h continuous operation. Training time had to be limited to 10 days. The pattern was then 2 days of baseline measures with up to 7 h sleep a night, 65 h of continuous operation, unlimited recovery sleep, and 2 further days (up to 7 h sleep a night) for recovery measures.

It was anticipated that in operational use the GCS would have to move to a new location fairly frequently. We simulated this schedule, while endeavouring to provide a link to more basic laboratory research, by alternating 5-h sessions on the GCS with 5-h sessions in our adjacent laboratory throughout the period, with 1-h meal breaks in between. For each session on the GCS, crews were confronted with a new area of terrain (photograph) and a fresh set of tasks.

Measures

(i) *System measures*

The subjects' work was complicated but in essence consisted of a series of tasks in which they were required, as a team, to fly the air vehicle tactically from a launch point to the target area while making various observations such as target detections and reporting on these. All keystroke data were logged, and from this several measures were derived, including throughput (tasks per hour), time on task, timeliness, and navigational accuracy. Each task was accorded one of three priorities (P1, P2, P3), the team being told that the higher priority tasks should take precedence. Although all these measures are relevant to the operation of such a system in real life, only a selection taken from the keystroke data can be presented at this time.

(ii) *Laboratory measures*

A battery of tests of cognitive function were used. Some tasks were selected because they represented standard laboratory paradigms, while others were chosen because they appeared to be analogous to aspects of the GCS operation. An example of the first category is the Five Choice Serial Reaction Time Task, but including an error detection measure. Examples of the second category are the Manikin Task (Benson & Gedye 1963), and a Data Entry Task. In the Manikin Task, subjects are presented with line drawings of a figure (the Manikin) holding a circle in one hand and a square in the other. One of these symbols is also present below the figure. The subject's task is to show in which hand (left or right) the Manikin is holding the symbol depicted below. The Manikin can be either facing or rear view and either upright or inverted. The task is therefore one of spatial decision-making. The Data Entry Task merely required the inputting of grid references via a keyboard. Discussion here will be confined to these three tasks.

(iii) *Subjective measures*

Mood was assessed during each laboratory session by using a computerised version of the UMACL, which was derived from Matthews (1983). This scale provides measures of arousal derived from Thayer's (1978) Dimension A (vigour–fatigue), Dimension B (tension–relaxation), plus Hedonic Tone (pleasure–displeasure) scales. Sleepiness was assessed every 2 h by using the Stanford Sleepiness Scale (Hoddes *et al.* 1973).

(iv) *Physiological measures*

Three physiological measures were taken, the first two being included primarily to provide additional data for other programmes. Saccadic eye movements were recorded during every laboratory session and analysed for sleep-loss effects by a componential technique (Green 1986); also, every 2 h throughout the trial, a 1 ml saliva sample was obtained for cortisol determination (Walker *et al.* 1978), and oral temperature was measured.

Results

(i) *System measures*

Systems measures were analysed by using Analysis of Variance (ANOVA); a full account will appear in papers by Rejman & Green (in preparation). In general, crew performance on the system held up very well throughout the trial. Thus the total mean number of tasks (P1, P2,

P3) accomplished per hour, a direct measure of crew–system throughput, remained relatively constant. However, when the task is broken down into the three priorities the results are as shown in figure 3.

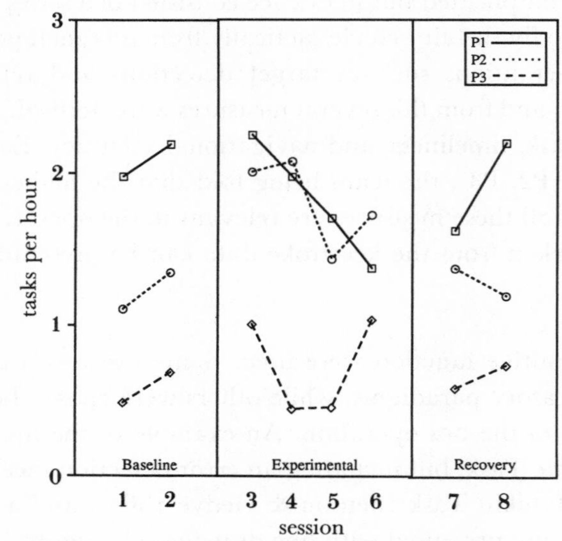

FIGURE 3. The effect of sleep loss on task throughput (tasks per hour) by priority.

Baseline sessions clearly show crews adhering to the priority scheme, with performance for the three priorities clearly separated ($F\,2, 8 = 47.5$, $p < 0.001$). This situation alters subtly under conditions of sleep loss. First, performance of P2 tasks is elevated beyond control levels to the extent that there is no significant difference between P1 and P2. Thereafter P1 and P2 tasks decline together as a function of sleep loss, with P3 task performance lower and less consistent ($F\,7, 28 = 6.36$, $p < 0.001$). However, main effects should be treated with caution because of the significant interaction ($F\,14, 56 = 2.71$, $p < 0.01$) between task priority and sleep loss. Recovery performance shows a virtual return to baseline levels for all three priorities. Subjects appeared to be attempting to keep up performance on both P1 and P2 tasks, but one result of such a combination was that P1 tasks suffered.

Despite these changes to the number of tasks accomplished, subjects did not appear to alter the mean time they devoted to each task once they had started it. As can be seen from figure 4 this measure changes very little (no significant differences) across the experiment, with, in general, P1 tasks receiving slightly more time than P2 tasks, which in turn receive more time than P3 tasks.

Finally, on timelinesss, figure 5 shows 'minutes late on task'. Here again performance can be seen to be governed by the priority scheme, with P1 tasks always being performed closer to the requested time than P2 and P3 tasks ($F\,2, 8 = 7.26$, $p < 0.05$). An exception to the priority pattern occurs towards the end of the experimental period when P3 takes precedence over P2. There was a significant interaction ($F\,14, 56 = 1.95$, $p < 0.05$) of priorities and sessions, although no significant differences were found between control, experimental, and recovery periods. A further blurring of the priority distinctions occurred in the last recovery session.

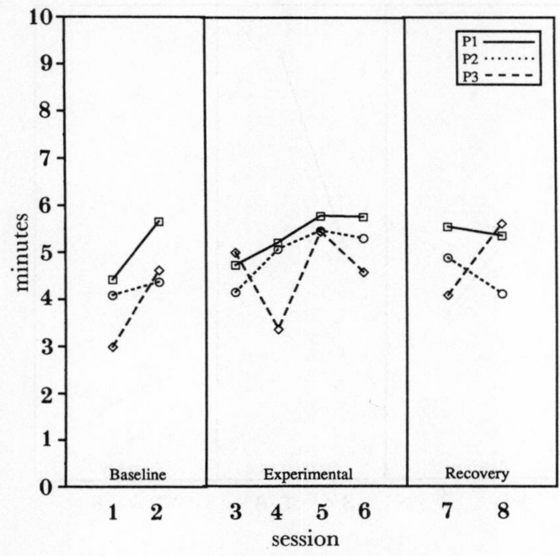

FIGURE 4. The effect of sleep loss on time spent on task by priority.

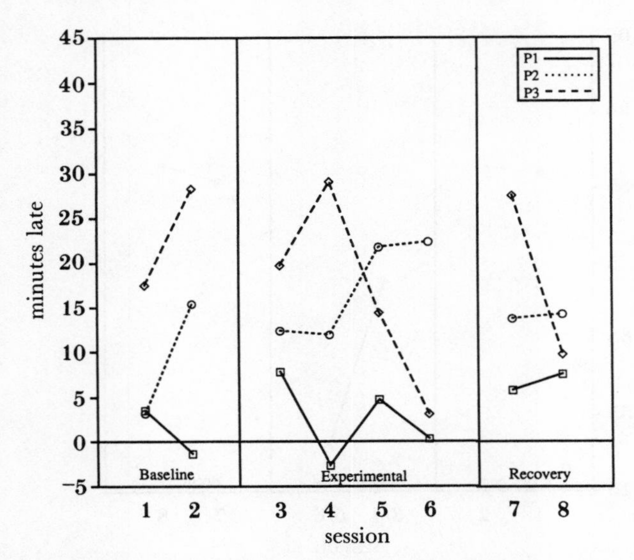

FIGURE 5. The effect of sleep loss on 'minutes late on task' by priority.

(*ii*) *Laboratory measures*

The cognitive tests were analysed using Analysis of Variance; a full account will appear in papers by Rejman & Green (in preparation). Data from the Five-Choice Serial Reaction Time Task (figures 6 & 7) are shown.

Figure 6 shows a significant increase (F 5, 85 = 15.21, $p < 0.001$) in choice reaction time with sleep loss, and also a practice effect across sessions. A further finding was that there was a significant decline (F 5, 85 = 34.83, $p < 0.001$) during sleep deprivation in error detection on this task (figure 7).

The Manikin Task also showed some evidence of learning throughout the trial on the latency score, a fact which appeared to offset the increase in reaction time with increasing sleep loss. However, errors did increase significantly during the sleep loss phase (F 6, 42 = 5.48,

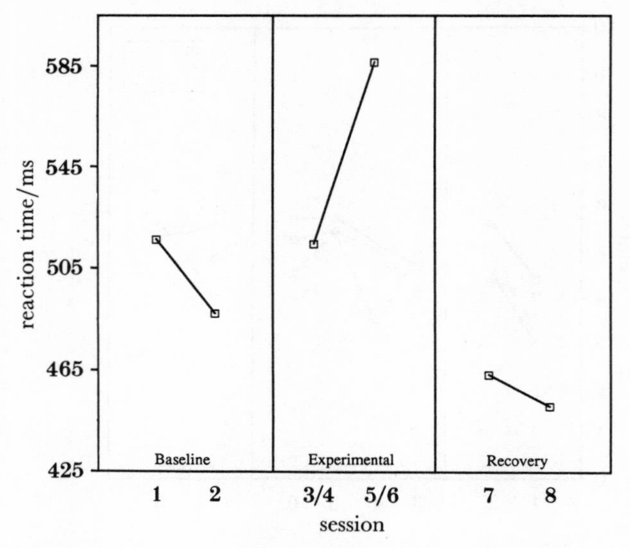

FIGURE 6. The effect of sleep loss on 5-choice reaction time.

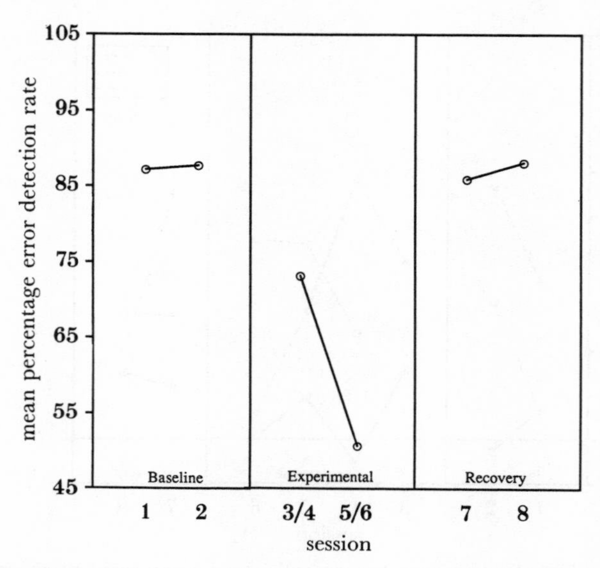

FIGURE 7. The effect of sleep loss on error detection rate in the 5-choice reaction time test.

$p < 0.001$). Thereafter they remained at this level into initial recovery, after which they returned to baseline levels. The Data Entry Task showed most effect on the latency measure with the time taken to input numbers significantly higher after sleep deprivation ($F\,6, 54 = 5.67$, $p < 0.01$).

Finally, although tentative at this stage, there may be some indication that aspects of performance are still vulnerable to errors after the recovery sleep that follows a period of deprivation. This needs further investigation but if borne out would have both theoretical and practical significance.

(iii) Subjective measures

ANOVA of the UMACL data showed the expected rise in self-reported fatigue with increasing sleep loss (F 5, 35 = 19.38, $p < 0.001$); it also showed an increase in tension (F 5, 35 = 10.53, $p < 0.001$) and a decrease in hedonic tone (F 5, 35 = 10.32, $p < 0.001$). Data from the Standford Sleepiness Scale are shown in figure 8, which show, as would be expected, increasing sleepiness with increasing loss of sleep (F 6, 736 = 35.35, $p < 0.001$).

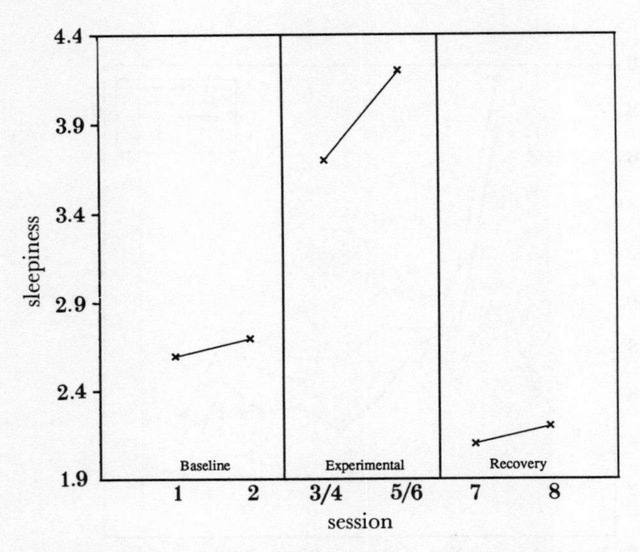

FIGURE 8. The effect of sleep loss on self-rated sleepiness.

(iv) Physiological measures

A typical result for the peak velocity component of the saccadic eye movement data is shown in figure 9, which shows the significant reduction(F 15, 75 = 9.12, $p < 0.001$) during sleep loss

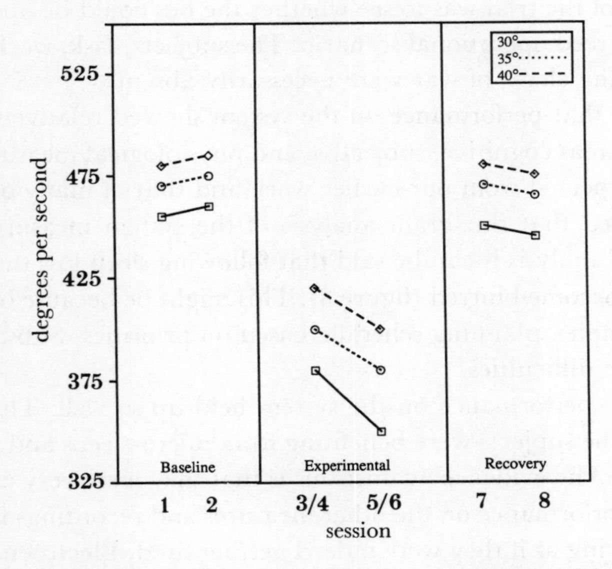

FIGURE 9. The effect of sleep loss on saccade peak velocity (degrees per second).

periods. With regard to cortisol, the baseline rhythm was as classically described by Kreiger (1979). Figure 10 shows that sleep loss rapidly perturbed the rhythm ($p < 0.01$) but with apparently unaltered total adrenocortical activity. After recovery sleep the normal rhythm was re-established but with evidence of increased adrenocortical activity as shown by a greater volume of secreted cortisol over the waking hours as compared with baseline ($p < 0.001$). The results for core temperature showed that as sleep loss increased there was a slight overall rise, with a forward phase shift. Recovery was marked by a rapid re-establishment of the normal circadian rhythm.

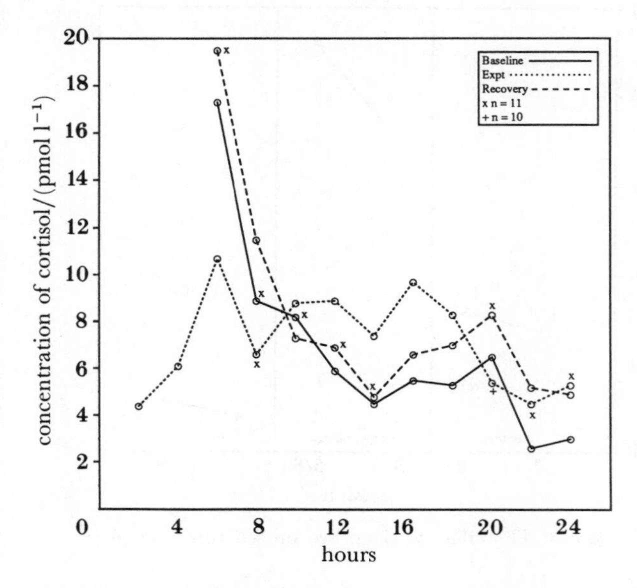

FIGURE 10. The effect of sleep loss on salivary cortisol secretions.

DISCUSSION

The primary purpose of the trial was to see whether the GCS could be operated satisfactorily in accordance with an agreed operational scenario. The subjects, task, workspace and scenario were realistic; fear and the chaos of war were necessarily absent.

Our main finding was that performance on the system showed relatively little change over the 65 h of sleep loss, whereas cognitive, subjective and physiological measures showed changes that would have been expected from our earlier work and that of many other researchers. It should, however, be noted that fine-grain analysis of the system measures is still going on. However, at this stage of analysis it can be said that following sleep loss the importance of the given priorities seems to become blurred (figure 3). This might be because of impairment of the ability to maintain a complex planning schedule based on priorities. Subsequent design work has tackled some of these difficulties.

We must now ask why performance on the system held up so well. There are three likely reasons. The first is that the subjects were benefiting from micro-sleeps and so were not as tired as we thought they were. The evidence against this is that they were very closely monitored at all times and that their performance on the other measures and recordings in the observers' log showed them to be behaving as if they were indeed getting tired. Electroencephalogram (EEG) recordings from earlier field trials, where the monitoring was slightly less intense, showed that subjects averaged 18 min unscheduled sleep on the second day of total sleep loss (Haslam 1982). However, it seems difficult to sustain the view that micro-sleeps were responsible for maintaining performance on the simulation, while clear decrements are revealed in the laboratory task performance. A second explanation is that the demands made on the subjects were not sufficiently taxing. Our only counter to this argument is that performance under both control and experimental conditions was still far from perfect, as the crew were kept in a state of permanent overload; however, under this sleep regime, it was sufficiently good to satisfy the Army that it would be adequate for their worst anticipated operational scenario.

The third explanation, and the one to which we subscribe, is that performance held up well because of the nature of the situation. This we believe to be a function of four factors.

(i) *Task design.* The human factors of the task had been optimized over several years by an iterative process with the future operators. Thus we had looked carefully at crew size, composition and interaction, task allocation, the working environment, situational awareness and man–hardware and man–software interactions, etc. Walker & Burkhardt (1965) have shown that performance decrement under stress on complex weapon systems is greater for those which were less well human engineered.

(ii) *The nature of the task.* The Ground Control Station was operated in 5-h bursts and the tasks were stimulating, varied, provided good feedback and required the active involvement of the subjects in most operations. Relatively few aspects involved either passive or repetitive action as we deliberately resisted the pressures to automate ('thereby removing human error'!) except to help overcome operator overload. Support for this approach comes from research, which shows that taking away functions from the operator makes him into a monitor and people do not make very good monitors (cf. Wickens 1984) in such circumstances.

(iii) *The motivation of the subjects.* Our subjects were very interested in their task and saw themselves as pioneers in a novel and important system, which they were likely to have to operate in the future. They were well trained professionals who reported that they welcomed the challenge of the trial.

(iv) *The group effect.* The system measures were, unlike those for cognitive, subjective, and physiological performance, based on group performance, and, as Morgan & Alluisi (1965), Davies & Tune (1969) and others have shown, group performance is often better than that of the individual. In addition, the system required the subjects to operate as a team, constantly interacting with each other to achieve group goals. One effect of this is that there were many times when it was observed that one crew member compensated for the flagging performance of a colleague who was going through a bad patch. The cohesion of service personnel is probably relevant here as this is their normal *modus operandi.* Certainly, when we tried to form a team of three civilians for a pilot study we could, even after prolonged training, only get them to function as three individuals.

Conclusions

The accumulated data from several laboratories on the effects of sustained operations on human performance did not allow behaviour in two archetypal military scenarios to be predicted with sufficient confidence, although they did provide a wealth of invaluable background material. A series of multidisciplinary trials did allow such predictions to be made and the data relating to the fairly severe field trials, augmented by information from many other sources, were translated into practical guidelines for military commanders.

The performance of three-man teams deteriorated as expected with increasing sleep loss when assessed singly by standard cognitive, subjective and physiological measures. But when assessed as a group on the system itself, such deterioration was far less, and it is concluded that the maintenance of performance (despite sleep loss) was a function of an ergonomically designed system, with active participation in a stimulating task by well-motivated subjects acting as a team.

We express our thanks to the Cranfield Institute of Technology, especially to Dr J. Ford and his staff, for software and hardware support and for assistance in running the trials. We are deeply grateful to Dr V. Schmit, Assistant Director, APRE, for much advice and assistance. Our thanks go also to Dr R. Edwards, APRE, who was responsible for the salivary cortisol study. Last, but not least, we thank the subjects, who showed such enthusiasm for the work.

References

Ainsworth, L. L. & Bishop, H. P. 1971 The effects of a 48-hour period of sustained field activity on tank crew performance. *Hum. Resour. Res. Org. Tech. Rep.*, no. 71–16.

Baddeley, A. D. 1972 Selective attention and performance in dangerous environments. *Br. J. Psychol.* **63**, 537–546.

Banks, J. H., Sternberg, J. J. & Farrell, J. P. 1970 Effects of continuous military operations on selected military tasks. *U. S. Army BESRL Tech. Res. Rep. no. TR*–1166.

Benson, A. J. & Gedye, J. L. 1963 Logical processes in the resolution of orientational conflict. *IAM Rep.* no. 259. RAF Institute of Aviation Medicine, Farnborough, Hants.

Berkun, M. M., Bialek, H. M., Kern, R. P. & Yagi, K. 1962 Experimental studies of psychological stress in man. *Psychol. Monogr.*, no. 15, vol. **76**.

Bourne, P. G. 1969 *The psychology and physiology of stress: with reference to special studies of the Vietnam war*. New York: Academic Press.

Davies, D. R. & Tune, G. S. 1970 *Human vigilance performance*. London: Staples.

Drucker, E. H., Cannon, L. D. & Ware, J. R. 1969 The effects of sleep deprivation on performance over a 48-hour period. *Hum. Resour. Res. Org. Tech. Rep.* no. 69–8.

Dudley, R. C., Brown, K. & Croton, L. M. 1974 The effect of five days' continuous day and night operations on Chieftain tank crew performance (Exercise Package Tour). Unpublished Ministry of Defence Report.

Green, S. 1986 Saccadic eye movement analysis as a measure of sleep loss effects on the central nervous system (CNS). In *Trends in ergonomics/human factors III* (ed. W. Karwowski), pp. 379–387. Amsterdam, Holland: Elsevier Science Publishers.

Haggard, D. F. 1970 Human Resources Research Organization studies in continuous operations. *HUMRRO Professional Paper*, no. 7–70.

Haslam, D. R. 1981 The military performance of soldiers in continuous operations: Exercise Early Call I and II. In *Biological rhythms and shift work* (ed. L. C. Johnson, D. I. Tepas, W. P. Colquhoun and M. J. Colligan), pp. 435–458. New York: Spectrum Publications.

Haslam, D. R. 1982 Sleep loss, recovery sleep and military performance. *Ergonomics* **25**, 163–178.

Haslam, D. R. 1983 The incentive effect and sleep deprivation. *Sleep* **6**, 362–368.

Haslam, D. R. 1985*a* Sleep deprivation and naps. *Behav. Res. Meth. Instru. Comp.* **17**, 46–54.

Haslam, D. R. 1985*b* Sustained operations and military performance. *Behav. Res. Meth. Instr. Comp.* **17**, 90–95.

Hartley, L. R., Morrison, D. & Arnold, P. 1989 Stress and skill. In *Acquisition and performance of cognitive skills* (ed. A. M. Colley and J. R. Beech), pp. 265–300. New York: John Wiley & Sons.

Hoddes, E., Zarcone, V., Smythe, H., Phillips, R. & Dement, W. C. 1973 Quantification of sleepiness: a new approach. *Psychophysiology*, **10**, 431–436.

Johnson, L. C. & Naitoh, P. 1974 The operational consequences of sleep deprivation and sleep deficit. *AGARDograph* no. 193. AGARD-AG-193.

Kreiger, D. T. 1979 Rhythms in CRF, ACTH and corticosteroids. In *Comprehensive endocrinology series 1: endocrine rhythms* (ed. D. T. Kreiger), pp. 123–141. New York: Raven Press.

Marshall, S. L. A. 1947 *Men against fire*. New York: Morrow.

Matthews, G. 1983 Personality, arousal states and intellectual performance. Ph.D. thesis. University of Cambridge.

Morgan, B. B. Jr & Alluisi, E. A. 1965 On the inferred independence of paired watchkeepers. *Psychon. Sci.* **2**, 161–162.

Murray, E. J. & Lubin, A. 1958 Body temperature and psychological ratings during sleep deprivation. *J. Exp. Psychol.* **56**, 271–273.

Rejman, M. H. & Ford, J. 1986 System simulation development as a tool in system design: a case history and a methodology. In *Trends in ergonomics/human factors* III (ed. W. Karwowski), pp. 163–171. Amsterdam, Holland: Elsevier Science Publishers.

Thayer, R. E. 1978 Towards a theory of multidimensional activation (arousal). *Motiv. Emot.* **2**, 1–33.

Walker, N. K. & Burkhardt, J. F. 1965 The combat effectiveness of various human operator controlled systems. In *Proceedings of the 17th US Military Operations Research Symposium.*

Walker, R. F., Riad-Fahmy, D. & Read, G. F. 1978 Adrenal status assessed by direct radio-immunoassay of cortisol in whole saliva or parotid fluid. *Clin. Chem.* **24**, 1460–1463.

Wickens, C. D. 1984 *Engineering psychology and human performance.* Colombus, Ohio: Merrill.

Wilkinson, R. T. 1962 Muscle tension during mental work under sleep deprivation. *J. exp. Psychol.* **64**, 565–571.

Wilkinson, R. T. 1964 Effects of up to 60 hours' sleep deprivation on different types of work. *Ergonomics* **7**, 175–186.

Williams, H. L. & Lubin, A. 1967 Speeded addition and sleep loss. *J. exp. Psychol.* **73**, 313–317.

Phil. Trans. R. Soc. Lond. B **327**, 543–553 (1990)
Printed in Great Britain

Circadian performance rhythms: some practical and theoretical implications

By S. Folkard

MRC/ESRC Social and Applied Psychology Unit, University of Sheffield, Sheffield S10 2TN, U.K.

Safety and productivity are low at night and this would appear to be because we are a diurnal species. This is reflected not only in our habitual sleep time, but also in our endogenous body clocks that, together with exogenous influences, such as the patterning of meals and activity, result in predictable circadian (24 h) rhythms in our physiological processes. Our performance capabilities also vary over the course of our waking period, with task demands affecting both the precise trend over the day, and the rate at which it adjusts to the changes in sleep timing occasioned by shift work. Studies designed to examine the reasons for this have shown that memory loaded performance may have a quite separate endogenous component to that responsible for more simple performance, suggesting that these two types of performance cannot be causally related. Furthermore, it would appear that the exogenous component of circadian rhythms may also differ across measures, and our attempts to model these endogenous and exogenous components have led us to re-examine the evidence on adjustment to night work.

Our findings suggest that shiftworkers merely 'stay up late' on the night shift, rather than adjust to it, and that this is responsible for the reduced safety at night. It would seem that in situations where safety is paramount, the only solution to these problems is the creation of a nocturnal sub-society that not only always works at night but also remains on a nocturnal routine on rest days.

Introduction

Many of the major industrial accidents involving human error have occurred at night. The Three Mile Island incident occurred at 04h00, Chernobyl at 01h23, Bhopal just after midnight and the Rhine chemical spillage in the early hours of the morning. This may, of course, be a coincidence, but the few studies that have obtained relatively continuous productivity and safety measures over the 24-hour day agree in suggesting that our performance capabilities are reduced at night (Folkard & Monk 1979). Thus, for example, single-vehicle accidents have been found to be three times more likely to occur between 21h00 and 09h00 than between 09h00 and 21h00, despite a considerably reduced traffic density. If this latter factor is taken into account, the relative probability of a single vehicle accident is some twelve times higher at night, and shows a clear peak between 03h00 and 06h00 (van Ouwerkerk 1987).

This problem of impaired performance at night would appear to stem largely from the fact that we have evolved as a diurnal species. This is reflected not only in the obvious fact that we habitually sleep at night, but also in the phase of the circadian (24 h) rhythms that are now known to occur in virtually all our physiological processes (Minors & Waterhouse 1981). These circadian rhythms are thought to reflect an evolutionary internalization of the pronounced 24 h changes in the physical environment. They enable species to anticipate these changes and hence have presumably been strengthened by natural selection.

Our measured circadian rhythms reflect the combined influences of an endogenous 'body clock' and a range of exogenous or 'masking' influences such as the habitual timing of sleep, activity, and meals (Wever 1979). Whether or not some of these exogenous influences are themselves controlled by a second, relatively exogenous oscillator or body clock, as Wever (1979) argues, is immaterial to this paper. Under normal sleep–wake conditions, the phase relation between the endogenous and exogenous components is relatively constant from day to day, resulting in predictable circadian variations in our physiological state.

Time of day effects in performance

In view of this circadian variation in physiological state it is hardly surprising that performance ability on various tasks has also been found to vary over the course of our normal waking period. At one time it was thought that performance on all tasks showed a similar pattern over the day, a view that echoed an early view of physiological circadian rhythms (J. Aschoff, personal communication). Performance efficiency was argued to parallel variations either in body temperature (Kleitman 1939) or, somewhat later, in basal arousal which was itself thought to largely parallel temperature (Colquhoun 1971). It is now clear, however, that this view was oversimplistic and that the nature of the variation in performance over the day depends, among other factors, on task demands.

In the case of a small number of performance tasks, fairly consistent trends in performance over the day have been observed, in several different studies (see Folkard (1983) for details of these). By expressing the value for each time of day in each study as a percentage of the mean value over the whole day for that study and then interpolating two-hourly readings, 'normative' time of day effects can be derived. Three such 'normative' trends are shown in figure 1 together with the standard deviation across the studies they are based on. This figure also shows 'normative' trends in body temperature and rated alertness derived in a similar manner. A number of points emerge from inspection of this figure.

First, despite the notorious 'noisiness' of performance measures, the standard deviations across studies are relatively small, showing a fair degree of consistency in the mean trends obtained. Secondly, the three different types of performance show very different trends across the day. Performance on simple serial search tasks, a task similar to proof-reading, improves over the day to reach a maximum at 20h00, whereas the immediate retention of information presented in short texts or films decreases over the day to reach a minimum at 20h00. Performance on tasks involving 'working memory', such as verbal reasoning and mental arithmetic, shows an intermediate trend. These differences in the trend over the day suggest that the short term or 'working' memory load involved in the performance of a task may be important in determining the trend over the day, a suggestion borne out by the results of other studies (see, for example, Folkard *et al.* 1976). Finally, there is a marked parallelism between performance on the simple serial search task and changes in body temperature, but no such parallelism with rated alertness.

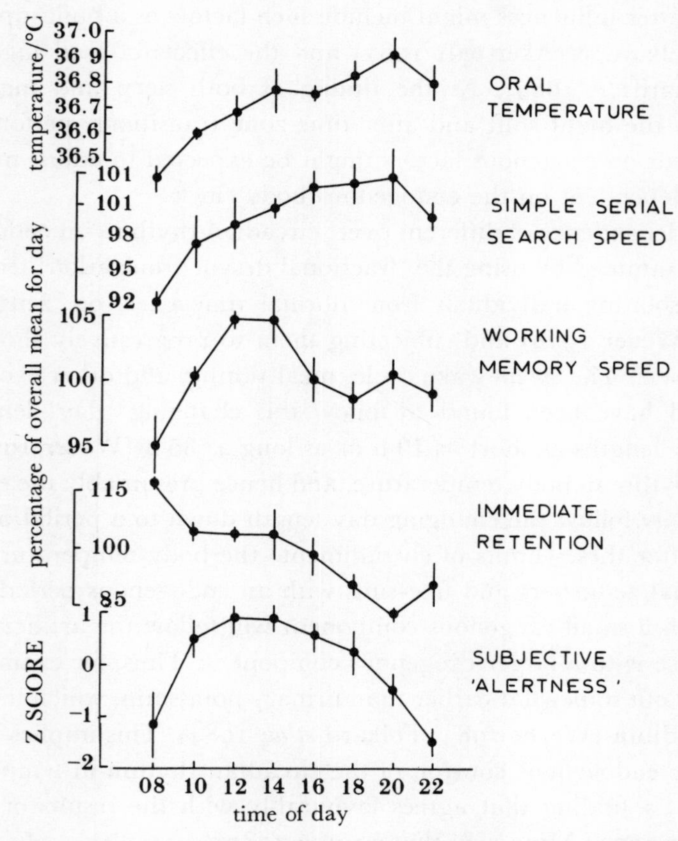

FIGURE 1. 'Normative' time of day effects in oral temperature, three different types of performance measure and rated alertness.

ADJUSTMENT TO PHASE-SHIFTS

These different trends in performance capability over the day could be interpreted as reflecting the combined effects of an underlying rhythm in basal arousal and task demand dependent differences in the 'optimal' level of arousal, that is, the Yerkes–Dodson law. However, such an explanation would predict that all performance rhythms should adjust at the same rate to one another to the change in sleep timing associated with shiftwork and rapid time-zone transitions. Although some support for this view has been claimed for a fairly restricted range of tasks (see Colquhoun 1971), there is evidence that the trend in 'working' memory performance may adjust to a change in sleep timing rather more quickly than that in either body temperature or performance on a serial search task (see, for example, Hughes & Folkard 1976). This has two important implications. First, it suggests that the optimal form of shift system may depend on the nature of the task being performed (Folkard & Monk 1979). Secondly, it clearly questions the validity of a simple arousal explanation, or any other single-factor explanation, of performance trends over the day.

The most parsimonious explanation of this task dependent difference in the rate of adjustment is that, like physiological rhythms, trends in performance over the waking period reflect both an endogenous body-clock component and the influence of various exogenous factors. These latter influences might include such factors as a build up of mental fatigue over time awake (Folkard & Akerstedt 1987) and the effects of food ingestion on performance (Craig & Richardson 1989). As the timing of both sleep and meals is changed almost immediately on the night shift and after time-zone transitions, performance trends that are largely dependent on exogenous factors might be expected to adjust more rapidly than those that are more dependent on the endogenous body clock.

The relative dependence of different overt circadian rhythms on endogenous and exogenous factors can be examined by using the 'fractional desynchronization' technique (Wever 1983). This involves isolating individuals from normal time cues, or 'zeitgebers', in a temporal isolation unit (Wever 1979) and subjecting them to progressively shortening or lengthening artificial zeitgebers. The sleep–wake cycle, meal timing, and other exogenous influences such as activity level have been found to follow this changing 'day' length with considerable accuracy to day lengths as short as 19 h or as long as 35 h (Wever 1979, 1983). In contrast, the circadian rhythm in body temperature, and hence presumably the endogenous body clock, will normally only follow the changing day length down to a period of about 22.5h or up to about 27.0h. After these 'limits of entrainment' the body temperature rhythm 'breaks out' from the artificial zeitgebers and free-runs with an endogenous period of 25 h.

Rhythms with a small exogenous component will follow the artificial zeitgebers to a lesser extent than those with a larger exogenous component. Thus, for example, body temperature typically breaks out somewhat earlier than urinary potassium, which in turn breaks out earlier than urinary sodium (Wever 1983; Folkard *et al.* 1984). This implies less exogenous control, and hence more endogenous control, of the circadian rhythm in temperature than of that in urinary sodium, a finding that agrees favourably with the results of other studies (see, for example, Wever 1979; Minors & Waterhouse 1981).

The use of this fractional desynchronization technique to examine the components of performance rhythms has produced some rather unexpected findings. In one series of studies (Folkard *et al.* 1983) individually isolated subjects were given performance tasks at the 'local time' equivalents of 06h30, 09h30, 12h30, 15h30, 18h30, 21h30 and 02h00. Subjects performed a simple serial search task and a verbal reasoning (working memory) task (Baddeley 1968) at each of these times, or different versions of a serial search and memory (SAM) task in which the memory load could be systematically varied from one (SAM-1) to five (SAM-5) items, for the 28 days duration of the study.

The circadian rhythm in performance on the simple serial search and SAM-1 tasks consistently followed that in body temperature in both shortening and lengthening studies. They had similar limits of entrainment and subsequently free-ran with indistinguishable periods. This contrasts sharply with the somewhat different limits found for the body temperature and urinary potassium rhythms (Folkard *et al.* 1984), despite both having a strong endogenous component (Minors & Waterhouse 1981), and implies that the rhythms in body temperature and serial search performance reflect a very similar mix of endogenous and exogenous influences.

About half the subjects run in this series of studies showed a roughly similar pattern of results in their verbal reasoning rhythm. In contrast, the remaining subjects showed a rapid phase advance of this rhythm, relative to both that in body temperature rhythm and their sleep–wake cycle, at some stage of the study. More detailed analyses of these sections of the total time series showed that their verbal reasoning rhythm was running independently with a 21 h period.

Further, all the subjects given the SAM-5 task either showed a 21 h period in their performance of it at some stage of the study or, in the case of (unpublished) shortening studies, their rhythms remained entrained to a period, of 21 h or less. In contrast their performance rhythm on an intermediate memory load version (SAM-3) showed an inconsistent pattern of results. This suggests that a 21 h period in performance may only occur if the subject's short-term or 'working' memory capacity is taxed by the task demands, and that individuals differ in the degree to which this capacity is taxed by tasks such as SAM-3 and verbal reasoning.

This finding of a 21 h period in memory-loaded performance is totally inexplicable in terms of the current view of the circadian system outlined above. It is, however, consistent with the more detailed analysis of the results of some earlier phase shift studies of performance rhythms (Folkard & Monk 1982). It implies the existence of a second, previously unidentified, endogenous body clock that has a free running period of 21 h.

The relations between psychological and physiological measures

The separation of the endogenous component of the rhythm in memory loaded performance from that in both simple serial search performance and body temperature also implies that there can be no causal relation between these functions. Indeed, Wever (1983) argued that any temporary separation of two overt rhythms must imply that there is no direct causal relation between the processes in which they occur, although, unlike a separation of the endogenous components, it clearly does not rule out their common control by different proportions of the same two underlying processes. As it is now clear that a single factor arousal theory is oversimplistic (Broadbent 1971) and that there must be a number of different arousal states (see, for example, Hockey & Hamilton 1983), there is a considerable interest in determining the relations between different physiological and performance measures.

The above results suggest that the circadian rhythmicity in different measures could be used to determine their functional relation or 'causal nexus'. Unfortunately, the number of measures included in these studies was limited by the high cost of assaying the large number of urine samples produced. There are, however, published studies that have obtained a greater range of measures and that can be re-analysed to explore this possibility. The first of these is a study in which subjects were sleep deprived and temporally isolated for 75 h (Froberg 1977). During this time exogenous influences were minimized by keeping activity level constant and giving identical three-hourly snacks. A range of performance and physiological measures were obtained at three hourly intervals, and the circadian rhythm in these would be expected to phase delay, or free-run, under these conditions (Aschoff *et al.* 1975). Our re-analysis involved estimating the best fitting period (that is, direction and extent of phase shift) for each variable and for each subject, and then examining the relation between these estimates (Folkard *et al.* 1986).

In view of the design of this study, no reliable rhythm could be detected for many of the subjects in the two measures with a strong exogenous component, namely, pulse rate and urinary noradrenalin. Our analysis was thus limited to measures with a relatively large endogenous component to their circadian rhythms. In most of the measures the endogenous circadian rhythm showed the expected tendency to phase delay. However, digit span, a classic measure of short-term memory capacity, had a mean best fitting period of 21.7 h, that is, it phase-advanced. This estimate was reliably shorter than that of any of the other measures taken ($p < 0.01$ in all cases) and concurs well with the results reported above. Further, and again consistent with our earlier findings, approximately half the subjects showed a mean best-fitting period of 21.4 h in their performance of a verbal reasoning task, while the remainder showed a phase delay.

To explore the relation between these estimates, principal components analyses with varimax rotation and hierarchical cluster analyses were performed. The results of these are shown in figure 2, together with the results of parallel analyses based on the mean level for each

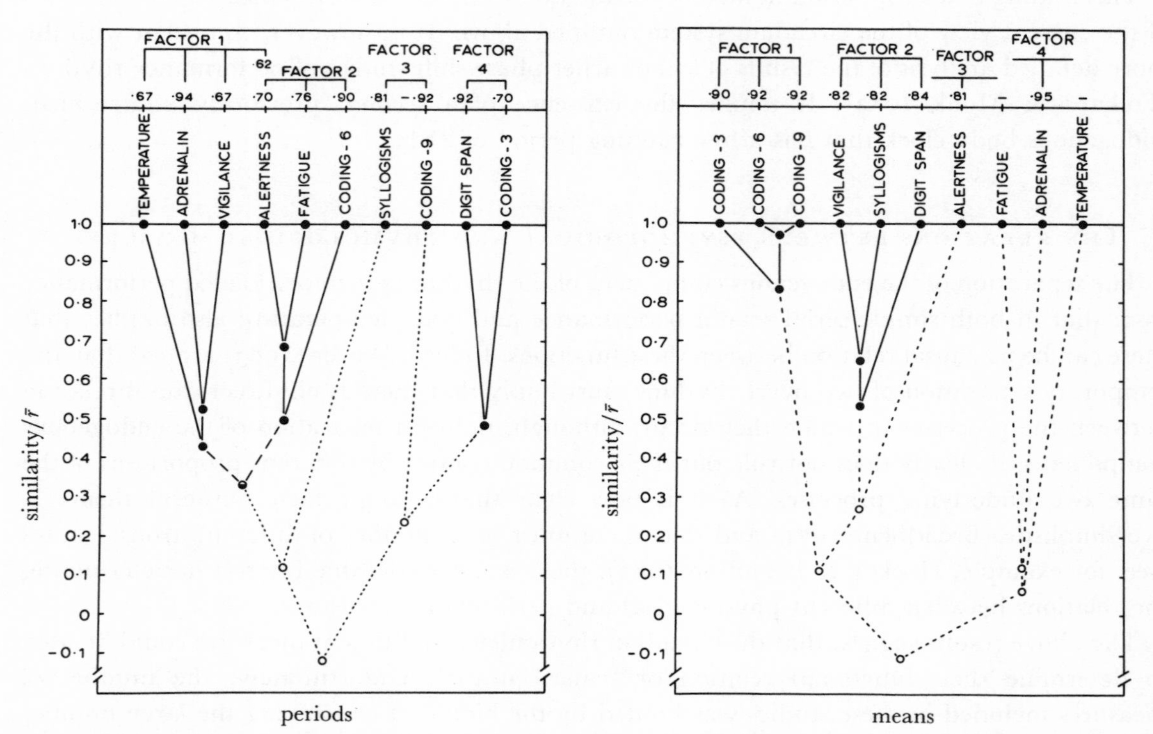

FIGURE 2. The relation between the measures as assessed by principal components (top panels) and hierarchical clustering (bottom panels) analyses based on the best fitting periods (*a*) and mean levels (*b*). Both figures show probable (solid lines), possible (dashed lines) and improbable (dotted lines) relations between the variables.

individual for each variable. This allows the present approach to be contrasted with that of Fleishman (see, for example, Fleishman & Quiantance 1984). The analyses based on best fitting period, that is, on the response of the endogenous circadian rhythm, showed a number of interesting results. First, it suggested that temperature, adrenalin and vigilance performance are probably related to one another, in agreement with the results of earlier studies (O'Hanlon 1965; Colquhoun 1971). Secondly, it suggested that this group of measures may also be related to a second group of related measures comprising ratings of alertness and fatigue and performance on a complex coding task. Finally, it confirmed that digit span performance showed no relation to temperature, but may have been related to performance on a more simple version of the coding task that could have been performed by using a high working memory load strategy. This pattern of findings contrasts sharply with that obtained by using Fleishman's approach that yielded two groups of performance measures, but no relation between these and either physiological measures or subjective ratings of affective state.

This pattern of results clearly confirms the potential usefulness of circadian rhythmicity in exploring the relation between variables. However, our attempts to extend this approach to studies of the phase shift imposed by eastward rapid time-zone transitions met with only limited success (Folkard *et al.* 1982). Our estimates of phase shift proved to be very noisy, presumably due to 'masking' by the exogenous factors that typically adjust almost immediately to such time-zone transitions.

Separating the endogenous and exogenous components

Our most recent efforts have thus been directed at trying to separate the endogenous and exogenous components of rhythms without the cumbersome and costly use of temporal isolation facilities. The reasons for this are twofold. First, it is clear that we can only examine the relation between endogenous circadian rhythms in, for example, rapid time-zone transition or shiftwork studies if we can somehow remove the exogenous component. Secondly, it would seem that the amplitude of circadian performance rhythms is insufficient in itself to account for the increased risk of accidents at night alluded to above. Rather, it would appear that there must be other factors associated with night work that contribute to impaired safety and productivity. The most obvious candidates are our natural inclination to sleep at night, and the reduced duration of sleeps taken between successive night shifts. Both these factors are influenced by our endogenous body clock.

Our first attempt to separate the endogenous and exogenous components of circadian rhythms was based on the results of a second series of fractional desynchronization studies (Folkard *et al.* 1984). The major finding from these studies was that the circadian rhythm in rated alertness not only broke out from the artificial zeitgebers, but did so before that in body temperature (Folkard *et al.* 1985). After this break out it free-ran through approximately 360° relative to the sleep–wake cycle and other exogenous influences. Thus the 'local time of day effect' after this break out could be argued to reflect the exogenous influences averaged over different phases of the endogenous component. Our more detailed analysis of this exogenous component showed that it could be described very accurately by two mathematical functions (Folkard & Akerstedt 1987). The more important, process 'S', is an exponential decrease in alertness (or increase in tiredness) over time awake, and is very similar to the hypothetical build up in sleep need postulated by Borbely (1982). The second, short lived, exponential function (process 'W') reflects a reduction from the extrapolated values of process 'S' on awakening, takes three hours to dissipate and is assumed to reflect the process of 'waking up'.

By subtracting this trend due to exogenous influences from the 'time of local day effect' before the break out, that is, when both exogenous and endogenous influences were present, we were also able to estimate the endogenous component. This was described very accurately by a 24 h cosine curve with an acrophase (peak) at 17h00. Thus the normal trend in alertness over the day could be accurately represented as the sum of three mathematical processes. By extrapolating these and summing their products it has also proved possible to account for the results of both prolonged sleep deprivation and phase shift studies with considerable accuracy (Folkard & Akerstedt 1989). Further, although based on alertness, this mathematical model has successfully been generalized to vigilance performance, and would appear to be capable of being generalized to other, non-memory loaded, performance measures (Spencer 1987). More recent and preliminary results suggest that we may also be able to extend this model to account for the recuperation of process 'S' during sleep, and the point at which individuals will spontaneously wake up. This would allow us to predict the level of sleep deprivation associated with any given shift system.

In parallel to this, we have estimated the endogenous and exogenous components of the body temperature rhythm based on published 'normative' trends obtained under different conditions (Folkard 1988, 1989). The advantages of using the temperature rhythm are: (i) that most published field studies of the adjustment of circadian rhythms to phase shifts have included body temperature measures; (ii) that laboratory studies of the circadian control of sleep have used the circadian temperature rhythm as a 'marker' for the endogenous body clock. Like that in alertness, our estimated endogenous component of the temperature rhythm approximated very well to a cosine curve with an acrophase of 17h00. However, the shape of the exogenous component was such that it could not easily be described in terms of simple mathematical functions, and hence could not be extrapolated. Nevertheless, these two components could be phase-shifted relative to one another, and then summed to simulate the trend for various shiftwork and rapid time-zone transition studies.

These estimates have been used to 'predict' the overt body temperature rhythm on the sixth successive night shift of the studies of Colquhoun *et al.* (Colquhoun 1971). This is the normal maximum number of night shifts before a change to a different shift or rest days (Kogi 1985), when any partial adjustment is likely to be lost. These night shift studies involved either an eight- or twelve-hour change in sleep timing, and the estimated exogenous component was thus phase delayed by these amounts and then summed with an unshifted endogenous component. In both cases, the resultant simulated temperature curves matched the actual curves very well, and considerably better than a partial adjustment of the whole curve. We (in collaboration with Drs Minors and Waterhouse) are currently extending this model to other phase-shift studies and are finding that an extension of this basic technique produces estimates of endogenous phase shift that are indistinguishable from those obtained by using costly and cumbersome 'constant routines' (Minors & Waterhouse 1981).

The conclusion to be drawn from this is that we may have grossly overestimated the adjustment of shiftworkers' endogenous body clocks to night work. Rather, over the course of a normal span of night shifts, shiftworkers would appear to simply 'stay up late' relative to their normally phased, endogenous clocks. As both the probability of falling asleep, and the subsequent sleep duration, are dependent on the phase of this body clock (for example, Zulley *et al.* 1981), this has two important implications for safety on the night shift. First, the individuals concerned will be trying to work when their ability to resist sleep is at its lowest ebb (Lavie 1986). Secondly, the duration of sleeps taken between successive night shifts is likely to be curtailed. Indeed, when translated onto 24-hour time, the body clock control of sleep duration in temporally isolated subjects cross-correlates highly ($r = 0.903$) with the time of day effect in shiftworkers' sleep duration (Folkard 1988). Consequently, a cumulative sleep debt may accrue over successive night shifts, which will probably exacerbate the night worker's natural tendency to fall asleep.

Discussion

At a theoretical level, these findings clearly suggest that the time of day effect in the performance of any given task will reflect a mix of endogenous and exogenous influences. Further, the endogenous component would itself appear to differ with task demands. In the case of simple serial search and vigilance performance it would seem to be the same as that responsible for urinary adrenalin, body temperature and probably rated alertness. The temporary separation of these latter two rhythms observed by Folkard *et al.* (1985) can now be attributed to their rather different exogenous components rather than to a more fundamental difference. In contrast, the endogenous component of performance on highly memory loaded tasks would appear to be totally independent to that contributing to these other rhythms. As yet, no physiological circadian rhythm has been identified that is controlled by this second endogenous (21 h) oscillator, but this may simply reflect on the very limited range of measures that have been taken.

One, admittedly speculative, interpretation of these results is that they reflect separate oscillatory control of the two hemispheres. Such separate control is suggested by Reinberg *et al.* (1988) who found the rhythms in right- and left-hand grip strength to desynchronize. In the present context, it seems possible that the memory loaded tasks may have placed greater reliance on left hemisphere activity, and indeed there is some evidence that it may be the verbal component of these tasks, rather than working memory load *per se*, which is important in determining the trend over the day. However, it should be noted that Reinberg *et al.* (1988) also found the period of the rhythm in right, but not left, hand grip strength to be correlated with that of body temperature, a finding that appears to contradict any simple interpretation of these results.

Unfortunately, it also seems probable that the exogenous component of a performance rhythm will vary across tasks, both in terms of its relative size and its shape. We already know that the exogenous components of the temperature and alertness rhythms differ, and it seems unlikely that those of different performance will not do so. We are thus faced with the possibility that similarities or differences between the trends in performance over the day may tell us little, if anything, about their underlying control (cf. Broadbent *et al.* 1989). Nor need differences in the rate of phase shift of overt rhythms help much in this respect as they could simply reflect differences in their exogenous components. It is only if one component can be removed, either experimentally or by modelling (see above), can inferences concerning causality be made based on the behaviour of the other component.

At a more practical level, the implications seem rather more straightforward. Reduced productivity and safety at night would appear to be due to the shiftworkers' relatively normally phased body temperature clocks. As a consequence, they will be trying to work when many of their performance capabilities are at a low ebb, when they are relatively unable to resist sleep, and when they may have accrued a cumulative sleep debt. This latter problem can be minimized by the use of rapidly rotating shift systems that involve a maximum of two or three successive night shifts, and which are generally socially acceptable. Such systems may also be beneficial for more memory-loaded jobs (Folkard & Monk 1979).

However, in situations where safety is paramount, such as in the control room of a nuclear power plant, more extreme measures may be required. The best solution here would appear to be to create a nocturnal sub-society that not only always works at night, but also remains on a nocturnal routine on rest days. It seems that only under these extreme conditions will the shiftworkers' body clocks totally adjust, and remain adjusted, to night work and hence safety be raised to daytime levels.

I thank my colleagues and collaborators for their contribution to the studies discussed in this paper, and without whom this paper would not have been possible. In particular W. Peter Colquhoun and Timothy H. Monk (Sussex); Torbjorn Akerstedt and Jan E. Froberg (Stockholm); Hans M. Wegmann (Cologne); Rutger A. Wever (Munich); David S. Minors and Jim M. Waterhouse (Manchester), and Mick Spencer (Farnborough).

References

Aschoff, J., Hoffman, K., Pohl, H. & Wever, R. A. 1975 Re-entrainment of circadian rhythms after phase shifts of the zeitgebar. *Chronobiologia*, **2**, 23–78.

Baddeley, A. D. 1968 A 3-min reasoning test based on grammatical transformation. *Psychon. Sci.* **10**, 341–342.

Borbely, A. A. 1982 A two process model of sleep regulation. *Hum. Neurobiol.* **1**, 195–204.

Broadbent, D. E. 1971 *Decision and stress*. London: Academic Press.

Broadbent, D. E., Broadbent, M. H. P., & Jones, J. L. 1989 Time of Day as an Instrument for the Analysis of Attention. *Eur. J. Cog. Psychol.* **1**, 69–94.

Colquhoun, W. P. 1971 Circadian variations in mental efficiency. In *Biological rhythms and human performance* (ed. W. P. Colquhoun), pp. 39–107, London: Academic Press.

Craig, A. & Richardson, E. 1989 Effects of Experimental and habitual lunch-size on performance, arousal, hunger and mood. *Int. Arch. Occup. Environ. Health* **61**, 313–319.

Fleishman, E. A. & Quiantance, M. K. 1984 *Taxonomies of human performance*. New York: Academic Press.

Folkard, S. 1983 Diurnal Variation. In *Stress and fatigue in human performance* (ed. G. R. J. Hockey), pp. 245–271. Chichester: John Wiley & Sons.

Folkard, S. 1988 Circadian rhythms and shiftwork: Adjustment or masking? In *Trends in chronobiology* (ed. W. Th. J. M. Hekkens, G. A. Kerkhof, & W. J. Rietveld), pp. 173–182. Oxford: Pergamon Press.

Folkard, S. 1989 The pragmatic approach to masking. *Chronobiol. Int.*, pp. 55–64.

Folkard, S. & Akerstedt, T. 1987 Towards a model for the prediction of alertness and/or fatigue on different sleep/wake schedules. In *Contemporary advances in shiftwork research: theoretical and practical aspects in the late eighties* (ed. A. Oginski, J. Pokorski & J. Rutenfranz), pp. 231–240. Krakow: Medical Academy.

Folkard, S. & Akerstedt, T. 1989 Towards the prediction of alertness on abnormal sleep/wake schedules. In *Vigilance and performance in automatized systems* (ed. A. Coblentz), pp. 287–296. Dordrecht: Kluwer.

Folkard, S., Hume, K. I., Minors, D. S., Waterhouse, J. M. & Watson, F. L. 1985 Independence of the circadian rhythm in alertness from the sleep/wake cycle. *Nature, Lond.* **313**, 678–679.

Folkard, S., Knauth, P., Monk, T. H. & Rutenfranz, J. 1976 The effect of memory load on the circadian variation in performance efficiency under a rapidly rotating shift system. *Ergonomics* **19**, 479–488.

Folkard, S., Marks, M. & Froberg, J. E. 1986 Towards a causal nexus of human psychophysiological variables based on their circadian rhythmicity. *Rev. Physiol.* **16**, 1–9.

Folkard, S., Minors, D. S. & Waterhouse, J. M. 1984 Is there more than one circadian clock in humans? Evidence from fractional desynchronization studies. *J. Physiol.* **357**, 341–356.

Folkard, S. & Monk, T. H. 1979 Shiftwork and performance. *Hum. Fact.* **21**, 483–492.

Folkard, S. & Monk, T. H. 1982 Circadian rhythms in performance – one or more oscillators? In *Psychophysiology 1980 – memory, motivation and event-related potentials in mental operations* (ed. R. Sinz & M. R. Rosenzweig), pp. 541–548. Jena: VEB Gustav Fischer Verlag; Amsterdam: Elsevier Biomedical Press.

Folkard, S., Wegman, H. M. & Klein, K. E. 1982 Re-entrainment by partition following eastward (phase-advance) rapid time zone transitions. IB 316–82–03. DFVLR-Institut fur Flugmedizin: Linder Hohe, Koln.

Folkard, S., Wever, R. & Wildgruber, Ch. M. 1983 Multioscillatory control of circadian rhythms in human performance. *Nature, Lond.* **305**, 223–226.

Froberg, J. E. 1977 Twenty-four-hour patterns in human performance, subjective and physiological variables and differences between morning and evening active subjects. *Biol. Psychol.* **5**, 119–134.

Hockey, R. & Hamilton, P. 1983 The cognitive patterning of stress states. In *Stress and fatigue in human performance* (ed. R. Hockey), pp. 331–362. Chichester: Wiley.

Hughes, D. G. & Folkard, S. 1976 Adaptation to an 8-hour shift in living routine by members of a socially isolated community. *Nature, Lond.* **264**, 432–434.

Kleitman, N. 1939 (revised 1963) *Sleep and wakefulness*. Chicago: University of Chicago Press.

Kogi, K. 1985 Introduction to the problems of shiftwork. 1985, In *Hours of work: temporal factors in work-scheduling* (ed. S. Folkard, & T. H. Monk), pp. 165–184. Chichester: John Wiley & Sons.

Lavie, P. 1986 Ultrashort sleep-waking schedule. III. 'Gates' and 'forbidden zones' for sleep. *Electroenceph. clin. Neurophysiol.* **63**, 414–425.

Minors, D. S. & Waterhouse, J. M. 1981 *Circadian rhythms and the human*. Bristol: Wright-PSG.

O'Hanlon, J. 1965 Adrenalin and noradrenalin: relation to performance in a visual vigilance task. *Science, Wash.* **150**, 507–509.

Ouwerkerk, van F. 1987 Relationships between road transport working conditions, fatigue, health and traffic safety. VK 87-01. Rijksuniversiteit Groningen: Traffic Research Centre.

Reinbeerg, A., Motohashi, Y., Bourdeleau, P., Andlauer, P., Levi, F., & Bicakova-Rocher, A. 1988 Alteration of period and amplitude of circadian rhythms in shift workers. *Eur. J. appl. Physiol.* **57**, 15–25.

Spencer, M. B. 1987 The influence of irregularity of rest and activity on performance: a model based on time since sleep and time of day. *Ergonomics* **30**, 1275–1286.

Wever, R. A. 1979 *The circadian system of man: results of experiments under temporal isolation.* New York: Springer.

Wever, R. A. 1983 Fractional desynchronization of human circadian rhythms: a method for evaluating entrainment limits and function interdependencies. *Pflügers Arch.* **396**, 128–137.

Zulley, J., Wever, R. A., & Aschoff, J. 1981 The dependence of onset and duration of sleep on the circadian rhythm of rectal temperature. *Pflügers Arch.* **391**, 314–318.

Discussion

S. D. ROSEN (*Charing Cross Hospital, London, U.K.*). Dr Folkard has very convincingly shown the impairment of performance that occurs in shift workers, and the role of the body's clocks and exogenous cues in maintaining the stability of cycling internal systems. I feel that it must again be stated that besides the effect upon industry of impairment of performance of workers, shift work is frequently very injurious to their health because of the cumulative sleep deprivation. Long-term follow-up of the adverse effects on many aspects of health of chronic sleep deprivation have been demonstrated in very large numbers of individuals followed by the U.S. Cancer Registry. Also, our own (smaller) experience has noted an excess of cardiovascular morbidity in shift workers.

Secondly, it would be of great interest to assess a possible connection between perturbations of the temperature–adrenaline–vigilance cycle and the known circadian rhythms of angina, platelet aggregation and myocardial infarction all of which have been related to cyclical catecholamine excesses.

The above, coupled with Dr Folkard's suggestion of separate biological clocks in the two cerebral hemispheres, is of particular interest to us in the light of collaborative research between our own department, and the Charing Cross Neuropsychophysiology Laboratory, showing hemispheric asymmetry and disturbed left hemispheric function in patients with cardiovascular disorders.

Phil. Trans. R. Soc. Lond. B **327**, 555–567 (1990)
Printed in Great Britain

Automating assistance for safety critical decisions

By J. Fox

Imperial Cancer Research Fund Laboratories, Lincoln's Inn Fields, London WC2A 3PX, U.K.

Computer systems are increasingly being introduced to assist in decision making, including hazardous decision making. To ensure effective assistance, decision procedures should be theoretically sound, flexible in operation (particularly in unpredictable environments) and effectively accountable to human supervisors and auditors. Strengths and weaknesses of classical statistical decision models are discussed from these perspectives. It is argued that more can be learned from human decision behaviour than has traditionally been assumed, and this motivates the concept of a symbolic decision procedure (SDP). The SDP is defined, described in terms of first-order (predicate) logic, and its use illustrated in a decision support system for medicine. We point out that the classical numerical decision procedure is a special case of a generalized symbolic procedure, and discuss the potential for rigorous formalization of the latter. We conclude that symbolic decision procedures may meet requirements for assisting human operators in hazardous situations more satisfactorily than classical decision procedures.

Introduction

A decision is a conscious choice between at least two possible courses of action
(Castles *et al.* 1971)

In recent years computer based decision support systems (DSSs) have become increasingly used in fields like medicine and industrial process-control where decisions may have critical consequences for safety. Performance demands may also be severe. For example, action may be needed urgently; available information may be imperfect or incomplete; even the options open to the decision maker may be incompletely worked out. In parallel with the development of DSSs, and in particular the appearance of 'expert systems', concern has been growing about the potential for catastrophic errors created by these systems and, worse, the potential for catastrophes whose causes cannot be established.

Concern for the risks associated with expert systems is now so strong that it has spilled over into public discussion, such as a television broadcast in which American and British practitioners and critics argue the dangers of using expert systems in medical, military and similar applications†, and there have been calls for restrictions on the deployment of unsupervised or autonomous systems in safety critical situations (*The Boden Report* 1989). These concerns have implications for the design criteria of DSSs and they fall into two main categories:

(a) Performance issues

1. The decision procedure used by the DSS must perform well (make or recommend good decisions) even in the face of degraded data.

† *Electric Avenue*, B.B.C. 1989

2. It should be able to support a wide range of decision types where necessary.

3. Respond flexibly and appropriately should the circumstances requiring a decision change.

(b) Responsibility issues

1. If decisions lead to errors it must be possible to establish the reasons for those errors.

2. Where it is practical and appropriate provision should be made for a skilled human supervisor to exercise overriding control.

The central requirements for meeting these demands are that we have a comprehensive and sound yet intelligible decision procedure. Theoretical soundness has been the preserve of classical statistical decision theory. However, I argue in what follows that classical procedures may be unacceptably inflexible in the face of unanticipated events, and therefore require close supervision, but are relatively unintelligible to human users. Decision technology has been developing rapidly in recent years, with expert systems trying to address the problems of flexibility and accountability by using techniques from artificial intelligence (AI), but the growth of decision theory has not remained in step, with expert systems frequently being developed in a worryingly ad hoc way (Fox 1988).

This paper suggests an approach to resolving these problems, which takes some inspiration from observations of human decision behaviour but can be formalized by using logical methods (specifically the first-order predicate calculus). First, a brief review of classical statistical decision theory is given.

Classical decision theory

Perhaps the most forthright statement of what should now be regarded as the classical theory of decision making is due to Lindley (1985):

> ... there is essentially only one way to reach a decision sensibly. First, the uncertainties present in the situation must be quantified in terms of values called probabilities. Second, the various consequences of the courses of action must be similarly described in terms of utilities. Third that decision must be taken which is expected – on the basis of the calculated probabilities – to give the greatest utility. The force of 'must', used in three places there, is simply that any deviation from the precepts is liable to lead the decision maker into procedures which are demonstrably absurd ...

There have been innumerable studies using this expected utility (EU) approach in designing DSSS for medical diagnosis and treatment (de Dombal 1972; Schwarz & Griffin 1986), business and politics (see, for example, Hill *et al.* 1979) and many other areas. How does EU theory meet the above requirements? Experience has shown that when Lindley's criteria are satisfied EU procedures frequently perform well and degrade gracefully if the quality or availability of relevant information falls (requirement 1 above).

Unfortunately, however, many practical decisions do not make it easy to meet Lindley's requirements. For example, classical decision procedures require that prior and conditional probability parameters for all decision options and relevant indicators are unambiguous and available, which they are frequently not. The notion of utility is problematic because costs and benefits are subjective, individualistic and difficult to quantify. The validity of measuring utilities of certain outcomes (such as 'quality of life' after surgery) is highly controversial.

Arguably the most serious concern about the numerical framework, however, is that it is a substantially incomplete account of what it means to make a decision. Classical procedures do not specify how we determine that a decision is required, for example, nor how to identify alternative decision options, nor how to select and manage data acquisition and other processes. In short, classical procedures have not been developed to 'respond flexibly and appropriately should circumstances change' (requirement 3).

Human decision makers, on the other hand, clearly are capable of flexibly organizing and reorganizing their actions. (Indeed human decision analysts are routinely required to 'structure' a decision problem before mathematical techniques can be applied.) How do they do this? Human decision behaviour has been widely studied and modelled in psychology (see, for example, Broadbent 1971; Fischoff *et al.* 1981; Kahneman *et al.* 1982). Theoretical accounts of human judgement have often drawn on the classical analysis (see, for example, Hill *et al.* 1979; Hogarth 1980), but as we shall see human decision making may be based on principles that are rather different from those of subjective expected utility (SEU) theory.

Turning to the questions of accountability and audit we can see that classical theory leaves something to be desired here too. Current DSSs are still too primitive to be delegated responsibility for all aspects of decision structuring and decision making. Consequently designers try to keep a human decision maker 'in the loop' in situations involving risk, to monitor and supervise. Here the character of decision theory can be problematic, both because the mathematical terms are unfamiliar to the non-specialist and, even if familiar, the practical implications of sets of numbers may be hard to establish. Consequently this reduces understandability and scope for intervention by the supervisor.

The remainder of this paper is concerned with whether it is possible to develop a new kind of decision theory, which is inspired by human decision making to yield a procedure that is more flexible and intelligible than classical numerical procedures, but which can nevertheless be provided with sound theoretical foundations.

The rationality of human decision making

The basic assumption of classical decision theory is that it deals with rational choice. The decision maker is a rational individual who attempts to maximize the expected value of his or her actions, in the light of well-calibrated estimates of the likelihood of events and the costs and benefits associated with the outcomes of alternative actions. Many studies have shown that human decision makers fall somewhere below this standard of rationality. Reasons for this are not hard to find; our knowledge and use of quantitative parameters can be imprecise; our preferences may be inconsistent; our ability to recall and bear in mind all relevant parameters imperfect, and we are subject to lapses of attention, information overloading and various forms of physiological stress. It is easy to conclude, and often has been, that human decision making is irrational by comparison with normative procedures.

In my view the standard normative criteria of rationality may be too restrictive. I can hardly deny the above observations but analysis of human decision making has concentrated too much on its weaknesses rather than its strengths. As Shanteau (1987) remarks in an analysis of competent or expert decision making '... my emphasis has been on investigating factors which lead to competence in experts, as opposed to the usual emphasis on incompetence'. Shanteau identifies a number of characteristics of expert decision makers. First, he observes that experts know a lot about their field of expertise. They know what is relevant to a specific decision, they know what to attend to in a busy environment, and they know when to make exceptions to general rules. Secondly, and perhaps most compellingly for this discussion, experts know a lot about what they know and they can make decisions about their decisions (Fox 1984). They have good communication skills and abilities to articulate their decision processes. Furthermore, they know which decisions to make, and which not to; they can adapt to changing task conditions, and they are able to find novel solutions to problems.

Whatever their weaknesses, human decision makers have capacities that are needed for flexible adaptation to an unpredictable and dangerous world. The classical model, in contrast, seems to impose fundamental restrictions that make it difficult to match these human capacities.

(*a*) *Restrictions on flexibility*

The expected utility decision rule presupposes that the conditions requiring a decision are predetermined and static. This is quite unacceptable for an autonomous decision maker, which cannot assume that all circumstances have been enumerated in detail before commissioning. Consequently, rationality criteria must include the ability to be rationally flexible, including the ability to: (i) recognize that a decision is needed; (ii) identify the kind of decision it is; (iii) establish a strategy for making it; (iv) formulate the decision options; (v) revise any or all of the above in the light of new information.

A decision system should be capable of autonomously invoking and scheduling these processes as circumstances demand. Classical theory offers no guidance for developing the necessary techniques.

(*b*) *Restrictions on communication*

The availability of an external agent who takes responsibility for the trickier decisions cannot be assumed, but we still expect DSSs to be accountable to their supervisors, and society at large. Therefore, a high level of communication between human supervisors or auditors wishing to examine, and potentially to intervene in, any aspect of the decision process, must be achieved. In general, a decision maker or DSS needs to be able to reflect on the decision procedure, to be able to examine: (i) decision options (what choices exist); (ii) data (the information available that is relevant to a choice); (iii) assumptions (about viability of options, reliability of data etc); (iv) conclusions (in light of data and knowledge of the setting).

Reflective capabilities should extend to the decision process itself, including: (v) the goals of the decision (what is the decision supposed to achieve); (vi) the methods being pursued (what justifies the current strategy); (vii) characteristics of specific procedures (applicability, reliability, completeness etc.).

The decision process should be able to communicate the results of all such reflections for accounting and audit, and should permit intervention in any aspect of the decision process by a supervisor.

A 'rational' theory of decision making must acknowledge these requirements. Classical decision procedures may be optimal in the sense that they promise to maximize the expected benefits to the decision maker, but they must be viewed as unsatisfactory in other ways. The root cause is that the theory fails to provide a framework for fully autonomous decision making, or for audit and sharing responsibility with a human supervisor.

The fact that human decision makers show a degree of flexibility and articulacy, which is dramatically greater than we can achieve with computers that use numerical algorithms alone, suggests ways of improving on current techniques. The next section presents a brief discussion of the origins of these human capacities.

A computational view of human decision making

From a computational point of view EU theory makes a strong commitment to a very specific representational formalism (by using real numbers to encode uncertainty, costs and benefits etc.) and to a process formalism (the algorithmic application of arithmetic operations). From an analogous perspective the representations and processes of human decision making look rather different.

(*a*) *Representation*

Human decision making does not depend much on numbers. Its most prominent features are often said to be an emphasis on recognizing patterns of data that are strongly associated with stereotyped situations (for example, an elderly patient with unexplained weight loss could have cancer) and relations are qualitative and varied (A causes B; A suggests B; B is a side effect of *A*, and so on).

Skilled experts are frequently able to discuss how they make their decisions as well as to make them. For example, they may say what decisions are current, the reasons why they are being attempted, which decisions have been recently taken, what remains to be done, and so forth. We would like to represent the decision process in a way that allows comparable abilities and for this logical and symbolic representations seem promising. Symbolic representations can provide an explicit record of the events, processes, conclusions, justifications etc, from which reports of the decision process can be constructed as required.

(*b*) *Processes*

Classical theory requires the manipulation of numerical probability and utility coefficients whereas human decision making seems to be more strongly influenced by 'heuristic' processes (Tversky & Kahneman 1974). These can be viewed as approximations to a correct decision model, or rules-of-thumb, or pragmatic strategies for rapid decision making. It is widely considered that human decision heuristics only embody probability or utility information implicitly, if at all.

Heuristics can be interpreted as expressing a restricted kind of logical reasoning. For example, decision rules may be expressed in a special case, or propositional, logic; if the patient is elderly and has lost weight then cancer should be considered (Fox 1980). Human reasoning may also exploit 'first-order' rules. These are reminiscent of inference rules in the predicate calculus, which refer to classes rather than individuals, as in:

> If Patient presents with Complaint and possible cause of Complaint is life threatening then investigation of Patient for Complaint is necessary.

Here capitalized terms are logical variables that can be read to mean 'the class of all patients', 'the class of all complaints' etc. (Technically, the variables are universally quantified.)

Whether human problem-solving and decision making truly use propositional or predicate logic (as opposed to processing information in ways which merely resemble logical inference) is an ambiguous and controversial question. For present purposes, however, we merely claim inspiration from human performance without proposing to simulate psychological mechanisms. Just as probability theory provides a clear framework for EU theory, classical logics are formally well-defined, suggest profitable ways of analysing decision processes, and they can be used to clearly specify versatile and intelligible DSSs as shown here.

The oxford system of medicine†

The Oxford System of Medicine (OSM) is an information and decision support system which is intended to provide: (a), a comprehensive medical information service, and (b), assistance to medical practitioners making a wide range of patient management decisions (Glowinsky *et al.* 1989). Among the design objectives for the system are that it should assist with a comprehensive range of decision types (diagnosis, treatment, investigation, risk assessment and referral etc.) while permitting close monitoring and supervision of decision processes. The OSM is aimed principally at medicine, specifically general practice, but the decision process is intended to be application-independent.

Figure 1 shows the way in which a symbolic decision procedure is encoded in the OSM; the decision procedure used employs first-order logic to encode a general decision strategy,

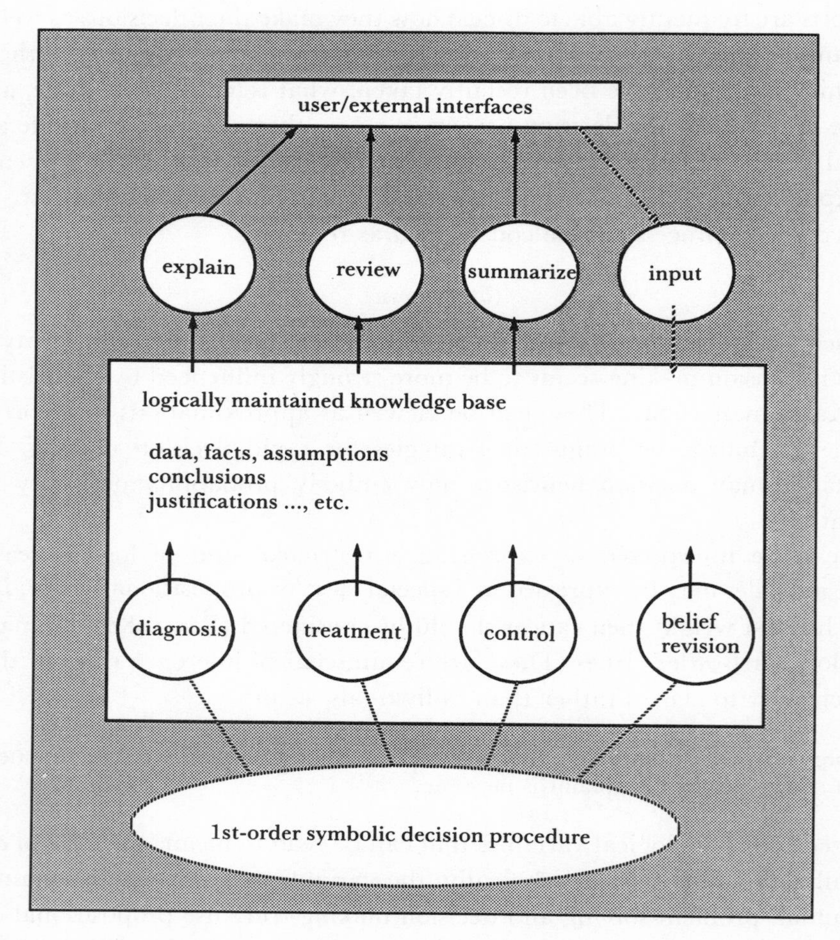

FIGURE 1. The functional organization of the Oxford System of Medicine. The symbolic decision procedure uses knowledge about specific decision tasks to interpret data input by the user, and to provide explanations, reviews, etc. Technically the knowledge base is logically maintained to automatically ensure consistency of beliefs using a truth maintenance system and logical knowledge of belief states (see text).

† The Oxford System of Medicine and some of the ideas in this paper were developed in collaboration with Andrzej Glowinski and Mike O'Neil. The early work on the design was partly supported by Oxford University Press.

incorporating methods for the following: (i) generating decision options; (ii) reasoning about the pros and cons of the options as data are acquired; (iii) grouping options together which are in some way inter-dependent; (iv) maintaining quantitative preference orderings over the (independent) options; (v) maintaining qualitative evaluative terms for the options.

The OSM incorporates a strategy for creating and executing specific decision tasks in response to user instructions. The details of specific medical decisions (diagnosis, treatment decisions, etc.) are represented as facts or propositions that are consulted by the decision procedure when a specific decision is required. These specify the methods and steps required to satisfy particular decision goals (eg. diagnosis, treatment). Strategies for controlling a decision process are also explicitly defined as part of the system's knowledge.

The abstraction of a general decision procedure from the specifics of medicine (or any other application) and the descriptions of particular kinds of decision and control strategies is central to satisfying some of the requirements discussed above. It contributes to formalization of the theory because essentials are clearly separated from the details of a particular decision; to accountability, similarly, because facilities for explanation and intervention can be designed, which are independent of the circumstances in which they may be required, but having access to all the system's knowledge of the application and knowledge of decision making strategies; to flexibility because the database of facts can be used for multiple purposes and, most importantly, because the system can reflect upon its own operation. The next section elaborates on and illustrates these ideas.

Symbolic decision procedures

'A symbolic decision procedure (SDP) is an explicit representation of the knowledge required to define, organize and make a decision, and is a logical abstraction from the qualitative and quantitative knowledge that is required for any specific application. A SDP may include a specification of when and how the procedure is to be executed.'

A SDP shares some characteristics of human decision making while preserving logical clarity. Human decision making is vulnerable to many influences that affect performance, so our aim is not to emulate human decision making in detail but to achieve a framework which embodies its desirable features within a rigorous computational framework. Features of the current design can be discussed in terms of representation and processing as introduced earlier.

(a) *Representation*

The symbolic decision procedure in the OSM represents patient data, medical facts, inferences made during decision making and their justifications, etc. explicitly as propositions, such as:

1. complaint('John Smith', weight-loss) *patient data*
2. causes(weight loss, cancer) *medical fact*
3. kinds(cancer, colon cancer)
4. possible(diagnosis('John Smith', colon cancer)) *a hypothesis*
5. confirmed-arguments('John Smith', diagnosis(colon cancer), support (causes (cancer, weight loss), kinds(cancer, colon cancer)) *a justified argument*

(b) Processes

These propositions are used by first-order logical inference rules. To put it another way, the rules, and hence the decision procedure, are generalized over classes of concept so that the definition is abstracted from the details of the (medical) application. Just as, say, Bayes' rule for the revision of probabilities specifies how to compute the posterior probability with the set of hypotheses, sources of evidence etc. supplied as numerical parameters to the procedure, a symbolic decision procedure is generalized as a first-order process that accepts qualitative rather than quantitative parameters. First-order processes for a symbolic decision procedure which deal with initiation and control of a decision process, reasoning about decision options, applying numerical decision procedures, achieving flexibility and robustness, and techniques for accountability and audit are now described.

Initiation and control of decision processes

The OSM is principally designed to assist with various medical decisions as required by a doctor. The decision procedures will therefore be initiated by explicit commands. More generally, however, symbolic decision procedures can initiate and control decision processes without supervision. For example, some sort of medical or industrial monitoring system may detect an abnormal parameter, decide to establish its cause and then go on to establish the appropriate treatment or action. If 'rapid weight loss' were observed, for example, and it could be logically established that cancer was among the possible causes of the weight loss (and cancer is of course pathological) then the rule would generate a requirement to take a diagnostic decision (Fox *et al.* (1989) provide more detail.). Recall Shanteau's observation that human experts know which decisions to take; they appear to be able to capture similar capabilities by logical reasoning about decisions and their preconditions.

Reasoning about decision options

Once a decision context has been established a number of basic operations will be carried out in that context, notably to reason about the options, relate options to each other (eg. one may be a special case of another), maintain preference orderings among them, and so forth (Fox *et al.* 1988). For example, we can define logic schemas that *argue* the pros and cons of a decision option. Each takes a number of symbolic terms as parameters, such as what the decision is, the set of decision options that are being considered and the set of arguments identified as supporting them. Argumentation is the process of constructing lines of reasoning and yields reasons for and against decision options for some specific case.

The decision procedure is modelled as a set of general schemas (that is, containing existentially or universally quantified variables) that operate over a database of propositions (see examples 1, 2 and 3 above). Proposition constants with the variables, and case-specific conclusions (such as 3 and 4) are deduced and added to the database. In what follows F, a finding, is any item of case data and O any decision option which might be considered for the case. A_T is an argument based on some theory T, about the truth of proposition P. The braces { } show sets, $\wedge$ is logical conjunction, the right arrow $\rightarrow$ logical implication and $\forall$, $\exists$ are universal and existential quantifiers.

A confirmed argument, conf_arg (C, P, A_T, S), is an assertion that a line of argument A_T about a proposition P applies to the current case, C. In the following scheme for conf_arg the sign, S, qualifies the argued proposition with one of: strengthening (+), weakening (−), confirming (++) or excluding (−−).

$$\forall F \forall C \cdot (\text{finding}(C, F) \wedge \text{provable}(F, P, A_T, S) \rightarrow \text{conf\_arg}(C, P, A_T, S)).$$

Where the predicate provable(Db, P, A_T, Q) means 'P may be proved from database Db by argument A_T from theory T with qualifier Q'.

The construction of arguments about decision options are particularly important; where there is a reason for considering the option O and no reason to exclude it then it is added to the option set for the case C, {option (C, O)}.

$$\exists A_{T_i} \cdot ((\text{conf\_arg}(C, O, A_{T_i}, S) \wedge S \in \{+, ++\} \wedge \sim \exists A_{T_j} \cdot (\text{conf\_arg}(C, O, A_{T_j}, --))) \rightarrow \text{option}(C, O)).$$

Here the predicate calculus has the advantage over wholly numerical methods of being able to express methods for introducing (and excluding) decision options incrementally, as information about the decision situation is accumulated.

Flexibility and robustness

The critical, indeed safety critical, question for DSSs is how we can provide them with the capability to take adaptive action in the absence of detailed knowledge of all the circumstances they may encounter. The difficulty with contemporary systems is that the facts or parameters required for all possible situations must be identified in advance. The contrast with people seems obvious; we are not immobilized by ignorance but can fall back on general knowledge of analogous circumstances and general problem solving strategies. A doctor can go back to first principles when faced with a difficult case, or apply general rules of good care while waiting for definitive findings.

One important resource available when facing new situations is theoretical knowledge. Theories can be introduced to interpret, hypothesize and predict circumstances that have not been encountered before because they capture important regularities of the world which (by definition) can be used to make predictions in unfamiliar circumstances. 'Common sense' understanding (of cause and effect; structure and organisation; function and purpose etc.) are included in the term theory. More sophisticated frameworks are also included, such as theories of biological and physical processes (which might be needed in medical decision making) and of course probability provides a well-developed theory for arguing about expectations. If we view theories as first-order inference procedures (that is rules that are generalized over significant classes of events) then theories can be formalized logically and used to argue hypothetically in circumstances where specific (explicit) medical knowledge is incomplete, uncertain or unobtainable, or to augment specific knowledge (see Fox *et al.* 1989) for more detail).

Symbolic procedures may be able to contribute to robustness in other ways. For example, skilled human problem solving and decision making can operate on two levels: the problem solving itself, and a reflective level which monitors and modifies the decision process as information is obtained. New information may suggest additions or corrections to the decision options, cast doubt on the evidence, suggest that the decision or problem needs to be reformulated, and so forth.

Numerical decision procedures as a special case of a SDP

Logical procedures for reasoning about the arguments for and against different decisions carry no information about the weight that should be attached to those arguments. However, although numerical representations may be relatively impoverished, we have acknowledged that classical decision procedures are well understood and they can offer quantitative precision. Symbolic decision procedures do not sacrifice the capability to use them where this is appropriate and practical. The logical elements can be extended with conventional probability and utility calculations. For example, we could include rules that revise probabilities in the light of findings. The schema below illustrates one way of specifying Bayes' rule for updating the probability of a hypothesis from the prior probability of the hypothesis and the finding and the conditional probability of F given H.

$$\forall F \forall H \cdot$$

$$(\text{probability}(H, P_H) \wedge \text{conditional\_probability}(F, H, P_{F|H}) \wedge \text{probability}(F, P_F) \wedge \text{bayes}(P_H, P_{H|F}, P_F, P_{H_{post}}) \rightarrow \text{probability}(H, P_{H_{post}})),$$

where $\text{bayes}(P_H, P_{H|F}, P_F, P_{H_{post}})$

is a function which returns a value for $P_{H_{post}}$ and is equivalent to

$$P_{H_{post}} = (P_H * P_{H|F})/P_F.$$

Calculation of expected utility, etc. can also be specified with such schemata (though note that some implementation details have been omitted for clarity).

Returning to the topic of robustness an ability to reflect on the decision procedure also offers advantages here. Mathematical operations, such as Bayesian revision, can be represented symbolically like any other concept. This permits the decision procedure to be extended to include schemata for reasoning about probability revision. Consider the case where the conditional probability table (of symptoms and diseases, say) is incomplete because hypotheses are generated in response to unusual combinations of observations. Under these circumstances the system should not ignore the missing data or, worse, fail to function. It should disable the Bayesian procedure, and continue to operate with, say, qualitative reasoning, while ensuring the problem is communicated to a supervisor or auditor.

Accountability and supervision

I have remarked several times that if decisions lead to errors it must be possible to establish the reasons for those errors, and that provision should be made for a supervisor to exert control. The OSM, for example, provides a range of reporting capabilities summarizing the decision options being considered at any moment, revising the direct and indirect arguments for any option, explaining why specific items of information may be relevant to the current decision and so on.

Good communication requires a rich vocabulary permitting the supervisor to ask questions like 'what diagnoses are presently suspected?', 'why is such and such implausible?' and to assert sentences like 'damage to heart is possible', 'treatment of breathlessness is urgent'. It may be practical to design such a language. A base proposition B such as 'diagnosis of Fred is cancer', can have distinct modalities such as possible *P*; unlikely *P*; suspected *P*. It is useful to compute certain modalities (such as 'possible diagnosis of Fred is cancer') or the general predicate mode(Decision, Case, Option, Mode)) from the pattern of support for the option (O) recorded in the database:

let Support = {conf_args(C, A_{T_i}, O, S)} then
option(C, O) ∧ provable(Support, A_{T_j}, O, M)} → mode(D, C, O, M).

Figure 2 shows part of an extended vocabulary for talking about beliefs; the modal terms are defined in terms of logical proof rules (definitions are given in the form of specific modal predicates for clarity).

Terms such as those in figure 2 have often been assumed to have a probablistic interpretation (for example, to say that 'cancer is possible' really means that cancer has a non-zero probability, or 'cancer is probable' that the probability is greater than 0.5 but less than 1.0, or some such.) These numerical interpretations raise many theoretical, empirical and technical problems, and I suspect that a logical interpretation is closer to their linguistic use (Fox 1987). This suspicion has received experimental support recently from an elegant series of experiments

If *arguments* can be identified which support an assertion, *P*, then *P* is *supported*:

arguments for *P* include argument→supported *P*.

P is *possible* if there is at least one supporting argument and *P* cannot be logically eliminated. (NB. This is an *a fortiori* definition. *A* proposition can also be *a priori possible* if the requirement for a supporting argument is relaxed.)

supported *P*, not eliminated *P*→possible *P*.

Arguments are arbitrary proofs over a body of knowledge. We may, for example, argue that a disease is a *possible explanation* of an observation by reasoning from physiology or other theories. Similarly we argue an assertion is *impossible* because it implies that an established theory is violated.

arguments against *P* include violation of Theory → impossible *P* (*and* hence → eliminated *P*).

Arguments can be counted to yield an unweighted measure of *support* (for a diagnosis, treatment, etc.):

possible *P*, number of arguments for *P* = Pos, number of arguments against *P* = Neg → support of *P* is Pos-Neg.

which can be translated into qualitative terms by testing simple arithmetic relations:

possible *P*, support of *P* is *SP*, *SP* > 0, not(support of *Q* is *SQ*, *SQ* > *SP*) → most supported *P*
possible *P*, support of *P* is Support, Support < 0 → dubious *P*.

We can define the concept of plausibility as 'a proposition is *plausible* if it is supported and there are no supporting facts or arguments that are themselves dubious':

supported *P*, not(*F* supports *P*, dubious *F*) → plausible *P*.

and *suspicion* rests on a slightly stronger condition:

plausible *P*, not(possible *P′*, support of *P′* > support of *P*) → suspected *P*.

Similarly, statements which have a conventional probability assignment can be tested to yield qualitative descriptors (strictly a partial ordering):

probability of *P* is Prob, not(probability of *Q* is ProbQ, ProbQ > Prob) → probable *P*
probability of *P* is ProbP, probability of *Q* is ProbQ, ProbQ > ProbP → improbable *P*.

Interestingly, probabilities can also have a second-order logical interpretation, as when we view a high probability as a kind of argument for an uncertain judgement:

probable *P* → arguments for *P* include highest probability of *P*.

So that *probability* arguments can, as intuition requires, yield judgements of *plausibility*.

FIGURE 2. Operators for natural uncertainty terms with logical scheme definitions. *P* is any proposition whose truth is uncertain. ("," is used here to represent logical conjunction.)

with native English speakers by Clark (1989) who was able to show by scaling analyses of some 50 belief terms that the underlying semantic space is multidimensional and highly organized.

The possibility that we can develop generalized logical vocabularies for urgency, importance, doubt and so on, which are closely allied to standard usages, may be of great value for human–computer communication. It should also be noted that a more expressive terminology will have value for decision making itself. For instance the concept of possibility plays a pivotal role in many decision schemata (Fox *et al.* 1989), and terms like plausible, suspected and so on may permit us to express more subtle yet intelligible conditions (Fox 1985). It can also be seen (see bottom of figure 2) that concepts such as 'probable' may be given a logical interpretation, both to express their meaning as a distinct set of logical conditions, and to exploit the concept, in a higher-order inference process. Just as Bayesian decision making can be viewed as a special case of a generalised symbolic decision procedure, so probability can be seen as a specialised uncertainty representation within a logical framework.

Soundness and formalization

One obvious feature of expert systems to date is that they do not have a well-defined decision procedure, but rather a collection of special-case rules for accumulating evidence. Consequently the decision strategy is implicitly distributed over the set of rules, which makes its behaviour uncertain and also a reflective capability is difficult to achieve. Furthermore, expert systems are frequently large and complex pieces of software; it is increasingly recognised that it is difficult to ensure the reliability of such software. Neither of these two characteristics is acceptable for hazardous decision making, but one can argue that the symbolic approach to decision making has potential for improving soundness of DSSs by formal analysis at a number of levels.

1. The semantics of the symbolic terms and propositions stored within the DSS potentially provide strong inter-constraints on the consistency of assertions: a man cannot have a gynaecological condition; symptoms do not cause diseases; a disease is not a possible option in a treatment decision, though it is of course a possible option for a diagnostic decision. These inter-constraints are useful for maintaining the integrity of the system as its knowledge increases, by manual revision or, potentially, by automatic methods of data and knowledge acquisition.

2. Constraints on the initiation and control of a decision derive from knowledge about situations and different types of decision. For instance, it does not make sense to try to make a decision among a number of possible decision options when the action in each case would be the same (as when it makes no sense to discriminate among a set of diagnoses when the treatment would be the same for all of them).

3. The clear definition of specific types of decision (such as diagnosis, treatment etc.) places logical constraints on how a knowledge base should be populated with propositions and theories that are required for each class of decisions.

4. Finally, it should be possible to formally specify a first-order decision procedure, to analyse its abstract properties separate from details of specific applications like medicine, and enforce appropriate class restrictions on variable assignments for example. It may even be possible to establish an a priori understanding of a procedure's performance characteristics, boundary conditions, failure modes and so forth. These are subjects of our further research.

Conclusion

The past few years have seen growing use of computer systems in hazardous settings, in both autonomous and supervized roles. The use of decision support systems, particularly expert systems, is attracting substantial concern. These concerns are well founded, but the growing calls for a moratorium on the introduction of such systems seem unrealistic. A different approach is to develop and formalize a symbolic decision theory that provides the basis for more versatile, more robust, more controllable and more accountable decision support systems than those based on classical expected utility theory, or first-generation expert systems. In this paper I have tried to show that human decision making, for all its vulnerability to error, is a source of inspiration for such a theory, and to illustrate how this inspiration can be turned into practical but formalizable techniques.

I thank Donald Broadbent, F.R.S., Dominic Clark and Mike Sternberg for helpful comments made during the preparation of this paper.

References

Boden, M. (chair) 1989 Benefits and risks of knowledge-based systems. *Report of Council for Science and Society*. Oxford University Press.

Broadbent, D. E. 1971 *Decision and stress*. London: Academic Press.

Castles, F. G., Murray, D. I. & Potter, D. C. (eds) 1971 *Decisions, organisations and society*. Harmondsworth: Penguin.

Clark, 1989 Psychological aspects of uncertainty and their implications for artificial intelligence. Ph.D. thesis, University of Wales Institute of Science and Technology.

de Dombal, T. 1979 Computers and the surgeon: a matter of decision. *Surg. Ann.* **11**, 33–57.

Fischhoff, B., Lichtenstein, S., Slovic, P., Derby, S. L. & Keeney, R. L. 1981 *Acceptable risk*. Cambridge University Press.

Fox, J. 1980 Making decisions under the influence of memory. *Psychol. Rev.* **87**, 190–211.

Fox, J. 1984 Formal and knowledge-based methods in decision technology *Acta. Psychologica*, **56**, 33–331. (Reprinted in 1988 *Professional judgement* (ed. J. Dowie & A. Eldstein). Cambridge University Press.

Fox, J. 1987 Making decisions under the influence of knowledge. In *Modelling cognition* (ed. P. Morris). London: J. Wiley.

Fox, J. 1989 Symbolic decision procedures for knowledge based systems. In *Handbook of knowledge engineering* (ed. H. Adeli). New York: Prentice-Hall.

Fox, J., Glowinski, A. & O'Neil, M. 1987 Towards a knowledge based information system for primary care. In *Information handling in general practice* (ed. R. H. Westcott & R. Jones). London: Croom Helm.

Fox, J., Glowinski, A., O'Neil, M. & Clark, D. A. 1988 Decision making from a logical point of view. In *Research and development in expert systems V* (ed. B. Kelly & A. Rector), pp. 160–175. Cambridge University Press.

Fox, J., Clark, D. A., Glowinski, A. J. & O'Neil, M. 1990 Using first-order logic to integrate qualitative reasoning and decision theory. *IEEE Trans. Syst. Man Cybern.* (In the press.)

Glowinski, A. J., O'Neil, M. & Fox, J. 1989 Design of a generic information system and its application to primary care. In *Proceedings of the Second European Conference on Artificial Intelligence in Medicine, Lecture Notes in Medical Informatics* (ed. J. Hunter), pp. 221–233. Berlin: Springer–Verlag, 1989.

Hill, P. H., Bedau, H. A., Chechile, R. A., Crochetiere, W. J., Kellerman, B. L., Ounjian, D., Pauker, S. G., Pauker, S. P. & Rubin, J. Z. 1979 *Making decisions*. Reading, Massachusetts: Addison-Wesley.

Hogarth, R. M. 1980 *Judgement and choice: the psychology of decision*. Chichester: John Wiley.

Kahneman, D., Slovic, P. & Tversky, A. (eds) *Judgement under uncertainty: heuristics and biases*. Cambridge University Press.

Lindley, D. V. 1985 *Making decisions*, 2nd edn. London: John Wiley.

Schwartz, S. & Griffin, T. 1986 *Medical thinking*. New York: Springer.

Shanteau, J. 1987 Psychological characteristics of expert decision makers. In *Expert judgement and expert systems NATO ASI Series* (ed. J. Mumpower), vol. F35.

Tversky, A. & Kahneman, D. 1974 Judgement under uncertainty: heuristics and biases. *Science, Wash.* **185**, 1124–1131.

Phil. Trans. R. Soc. Lond. B **327**, 569–576 (1990)
Printed in Great Britain

Temporal decision making in complex environments

By V. De Keyser

University of Liège, Department of Work Psychology, FAPSE-Bât. B 32, Sart Tilman, Liège 4000, Belgium

The basic hypothesis of the author is that under the influence of technological development and market pressure, situations take on temporal characteristics that are more and more difficult for the operator to control. The temporal strategies traditionally installed by the operator disappear, are transferred or transformed.

Far from counterbalancing these phenomena, the displays, as they are designed in the workplace, obliterate the temporal dimension. The errors that are seen to appear are the product of a mismatch between the characteristics of the situation and the operator's resources. Four mechanisms of time estimation are discussed. Field study results on temporal strategies, such as anticipation, assessment of a process evolution and planning adjustment are developed.

1. Introduction

(*a*) *The two faces of Janus*

Time, like the two faces of Janus, is at once a diagnostic tool and an element of complexity for man in dynamic environments. Today's research in continuous processes favours the complexity by emphasizing the existence of temporal errors and fixations: this is to say errors where the operator persists in a wrong course of action, or a wrong situation assessment, in the face of opportunities to revise (De Keyser & Woods 1989). This orientation masks to some extent the ways in which time contributes to the establishment of order in the environment: strategies and references used by the operator to improve his diagnosis are passed over in silence; but above all, insufficient emphasis is placed on the severe deficiencies of information displays in control rooms. Our basic hypothesis is that:

(i) under the influence of technological development and market pressure, situations take on temporal characteristics that are more and more difficult for the operator to control. The temporal strategies traditionally installed by the latter disappear, are transferred or transformed;

(ii) far from counterbalancing these phenomena, the displays, as they are designed in the workplace, obliterate the temporal dimension;

(iii) the errors that are seen to appear are the product of a mismatch between the characteristics of the situation and the operator's resources.

(*b*) *Profound changes in the control of continuous processes*

Over the past two decades, profound changes have occurred in the control of processes, and as a result they have modified not only the operator's role in the control room, his objectives and actions, but his knowledge as well. The principal modifications are shown below.

(i) *Modification at the automation level and of the system's computer structure*

A greater control of the parameters of industrial systems has allowed for a more stable automation and a larger computerization. The dynamic control of the process has gone further and further away from the operator, as well as the possibility of optimizing the parameters or diagnosing the incidents.

Many continuous processes now function according to the law of all or nothing; the system is stopped when it goes out of order and only the highly specialized teams for maintenance of quality control can intervene.

A greater emphasis is placed on recovery from incidents than on their prediction.

(ii) *Integration of the process into a larger and more highly interconnected system*

The processes have been widely integrated and made more complex. This renders them more vulnerable (Perrow 1984), and also more difficult to bring under control. As some of the fact always remain unknown to the operator, his understanding of the situation can never be complete.

(iii) *Modification of information displays*

The computerization of control rooms has brought with it the appearance of visual display screens giving a sequential information to the operator. These have replaced the mimic boards covering the walls of the control room that allowed a synoptic view of the system. Direct conversations by interphone between the operator and the specialized teams on the terrain often serve to supplement these displays.

(iv) *Different management methods*

New methods of management based on the clients' requirements have created the conditions for a different style of management and a 'just in time' planning. The disappearance, in most corporations, of stock, the emphasis on the economical challenge, and the rigorous time control are the new facets of the industrial game. The operator has more responsibility but is under more pressure.

However, if the operator's basic modes of time estimation are not modified, then they are integrated in different strategies. There is less control of the process, but more supervision of production. The operator keeps at hand only the transient conditions that are difficult to bring under control temporally. He manages a vast system in which he does not understand many of the indicators. With less diagnosis and more recovery, greater emphasis is placed on the prediction of the efficiency of actions than on an analysis of the origin of incidents. Based on the customer's requirements, a tight planning results that needs to adjust clock time with the process evolution.

2. Temporal estimation and characteristics of the work situation

(a) *The mechanisms*

Empirical studies (Decortis 1988; Van Daele 1988, 1989) and cognitive psychology experiments (Piaget 1946; Montangero 1979) provide the principal mechanisms that help to understand how people deal with time. Beside these mechanisms, the clock-time estimation intervenes as a metric.

(i) *Causal, or physical estimation*

The content of the elapsed events is used to evaluate time duration. The relationships established by the operator between velocity, space and time, correspond to the covariation of the physical system. This estimation mode is based on causality, and the time appears to the operator, as a time of action, that is incompressible if the given of the problem, speed and space, does not change. These mechanisms can only be used if the operator has a deep knowledge of the system, and if he has access to necessary information.

(ii) *Logical estimation*

The onset and the duration of the events are deduced from their relations with other events. The durations are perceived as intervals without causal content, only the beginnings and ends of intervals are meaningful for the subject. For instance, an operator could decide to start up a delicate procedure once the maintenance work has been finished. This mode of estimation is highly economic, it can be described as a 'temporal heuristic'. But it requires a sufficient saliency of cues, and the existence of other events for a relational structure to take place.

(iii) *Internal estimation*

Temporal regulations in man and animals have revealed the ability to learn rhythms and to estimate occurrences of events and durations that appear regularly. As the working situations are usually very familiar to the operators, they seem to gradually internalize more aspects of the environment with experience. But this subjective time estimation is the more exposed to possible temporal distortions.

These mechanisms of time estimation are used by the operator in varying degrees, and quite often he combines all of them in the same reasoning process. The preference given to one over another depends on the characteristics of the situation, the task, the information displays, and his own expertise. But another mode of time estimation, heterogeneous to the other ones, comes into play. It is the clock-time.

(iv) *Clock-time estimation*

Clock-time introduces into the situation a time that is exterior to both the action and the subject. It can be used as a convenient metric in some cases, but very often it creates a source of adjustment and tension. Most of the processes have to be run within temporal limits, and the operators have to match continuously the time of action and the clock-time.

(*b*) *The characteristics of the work situation*

De Keyser (1988) and Decortis *et al.* (1989) have described the work situations as a temporal structure of events, process events and human actions. The nature of the links, the occurrence of the events, their saliency are the temporal attributes. The new role of the continuous process operator is to integrate these events in parallel. We will focus the discussion on three temporal strategies of the operators: anticipation, assessment of the process evolution and planning adjustment.

3. Temporal strategies

(a) *Anticipation*

Since the 1960s, numerous studies carried out on continuous processes have brought to light a temporal strategy used quite frequently by operators: anticipation (Iosif 1968, 1969; Kortland & Kragt 1980). Anticipation is an oriented behaviour that permits the operator to get ahead of the event with a deftness and precision that often amazes the outside observer. Iosif best brought this strategy to light with his *in situ* analysis of operators of thermoelectric plants. He distinguishes two forms of anticipation:

(i) a probabilistic anticipation: the operator scans the parameters on the mimic board before they go out of order. He does this in a selective manner, 'as if', says Iosif, 'he has internalized the statistical structure of the board'; as if he had prior knowledge of the probability of appearance of breakdowns;

(ii) a functional anticipation: the operator predicts that a parameter will go out of order based on the variation of another one which is functionally connected to it.

In this research, the main indicator of the operator's performance was the relation between the time when the information was taken by the operator and the disturbance. To discuss the current validity of this work, two elements of the work situations that have been reported must be brought up. (*a*) Studies were carried out in control rooms where mimic boards were still being used. Iosif's work is based essentially on the associations the operator makes between the variations of the dials; on the emergence of covariation patterns with which he associates the states of the system. These patterns are not 'learned' in the framework of a training session, but rather are 'discovered' gradually by the operator. As the operator's practice accumulates, he assimilates rarer and differentiated patterns. (*b*) In these studies, the operator has kept, within his attribution, control of the process in case of an incident.

This research is already part of the past, for today anticipation has been profoundly transformed. The increasingly reduced action the operator takes on the process has brought about a decrease in consultation of the information displays, in spite of the large numbers remaining in the control room. As a result, the operators gain a much weaker knowledge of the process. A comparison of operators of the same technological steelmaking process, but at different levels of automation (De Keyser 1987) shows to what extent a high level of automation decreases the operator's knowledge of events. Not only do they misdiagnose incidents but, it is important to note, they do not attempt to anticipate them. This temporal strategy, which is brought up so often in the literature, has not completely disappeared; but has merely been transferred. The operator no longer has much to do with the process, but he has as an implicit task to manage a more vast system that encompasses the teams on the terrain, the transports and the relations with the forward and rear posts. This is where incidents can arise for him, and these are the incidents he tries to anticipate.

In anticipation, internalization and logical time estimation seem to be the main mechanisms. But the operator can no longer observe the actions of the team because of geographical constraints and he has only very poor instruments at his disposal for estimating the links that connect actions and events. Telephones and interphones are thus the preferred channels of this type of information – as described in field studies by Van Daele (1989). But the information he obtains in this way is far from precise and mentally costful, even if socially welcome.

(b) Assessment of the process evolution

Past research on the follow-up of evolutive processes especially emphasized the tasks of tracking. Today the action the operator takes on the process is being cut back; and apart from the manual control he exercises during transient conditions, the system evolves at least partially without him. Whenever the operator ceases to focus his attention on the anticipation of events, he continues to be very attentive to the supervision of the system changes. Indeed, it is during these moments that he makes decisions and coordinates actions.

In an electric powerplant, De Keyser *et al.* (1989) have studied the start-up process, comparing the knowledge and the performance of engineers and operators. They have recorded all the flow of behavioural traces, taking information from the displays, actions performed on the process, verbalizations, etc. during a 30-hour period.

The structure of these traces reflects the dynamics of the process, but also the expertise of the subjects. For both engineers and operators they reveal the existence of temporal envelopes. They are clearly shown by breaks, stops and more intense moments of information taking, which emphasize and punctuate the process evolution. We can distinguish two kinds of temporal envelopes.

(i) Temporal intervals, limited by indicators, which can be associated with relatively stable and known states of the process. All of the operator's attention is thus placed on the transitions and on the changes of state.

(ii) Temporal patterns, that is to say envelopes during which the parameters develop in a form known to the operator that he checks at regular intervals. They present interesting characteristics that distinguish them from intervals: these are personal constructions. They cannot be related directly, to specific states of the system. The breaks that the operator makes in the continuity of the system are his own. For him they correspond either to inflexion points of parameters, that he links to an action, or to critical values of variables which he must go through. But the operators and the engineers do not have the same patterns: those of the engineers seem to be more economic, more consistent with less checking, than those of the operators. The more practice the operators have, the better they are able to distinguish between subtle variations of state, by using appropriate cues. The prediction of the evolution of a process is thus internalized up to a certain point and monitored regularly by a checking. But without a causal model that would allow them to understand the evolution, rather than predicting, the operators anticipate certain indicators. Comparing the behaviour of pilots and expert monitors in transitory conditions, Amalberti *et al.* (1989) find the same results: they interpret the caution of the pilots and their numerous checkings as a fear about leaving a known temporal pattern. In a similar analysis, expert monitors prove to be much more audacious and consult the information displays less often.

The assessment of the process evolution seems to be characterized by the use of logical time estimation, building a nest of temporal envelopes to capture the dynamics of the environment. If these envelopes do not contain causal knowledge of the situation, if they are just shells, the operator does not seem able to really predict the future. Rather than predicting, he anticipates the apparition of indicators. And he tries to maintain himself in the limits of a known pattern. This is where we can see the advantage of intelligent aids that simulate an evolution of the process and the possible effects of certain actions. Such aids are still rare to this day. The limitations of these aids are bounded to the complexity of the situation. But certain situations are intrinsically complex and cannot be represented satisfactorily with mathematical models; in which case the operator prefers to rely on his own solid but hardly powerful experience, rather than to refer to an undependable mathematical model.

(c) Planning adjustment: when to act?

One of the most challenging tasks of an operator in a dynamic environment is to discover when he has to act. For the effects of an action can be totally different, if performed too early or too late. But the right time is not the clock-time: it depends upon the precise state of the process evolution. If the operator has a strong causal model of the process, he can manage himself by looking to cues he has related previously to specific changes of state. But the displays are rarely designed to provide this information in a form suitable for the operator. Even gradients, so useful to control the dynamic evolution of a plant, are usually not available on screens. When the operator has to intervene, not on the process but in the management of the system, the task is far more complex: the actions of the teams on the floor are not observable.

Very often the problem seems to be resolved by the existence of an organizational planning, which fixes the temporal structure of the system, and distributes the roles. Currently, new forms of management accentuate this tendency, providing the operator a computerized model. But the model does not include the incidents and the specific conditions of a given situation. In a continuous casting, Van Daele (1989) shows that 67 % of the planning programs estimated by the computer were corrected by the operators before being started. The corrections are not inversions of the sequence order of actions, rather, they are contractions or dilations of durations so as to take into account in advance the possible incidents of the system. They are, however, limited and concern a single post; where, according to the operators, the technical risk connected to the modification of duration was lower.

If planning requires a clear temporal limit, the operator must combine, in a causal manner, the metric of the clock with definite durations of actions. All of this expertise consists of estimating the possible adjustments between these two times, minimizing the risks, and taking into account the causal constraints of the system. It is clear that he hesitates to change the order of actions, or to cancel some of them. He does not seem to adopt opportunistic strategies, such as described by Hayes-Roth & Hayes-Roth (1979), may be because the mental workload entailed by this mode of planning is high (Valax 1985). But we can bring up another explanation. Many authors have represented the form of knowledge included in planning as schemata. But unless the operator has experimented himself with the effect of his actions, he is rarely certain about the links he establishes between events. Are these sequential links, or links of causality? In highly integrated industrial systems, any change has repercussions which spread. The operator acts very little; but when he does act, he limits his intervention to what he can keep under control.

Choosing the right moment to act, adjusting a fixed plan to the dynamics of the situation, are highly complex strategies. They often require a deep knowledge of the situation including the causal links between events and actions. The introduction of the clock-time in this game, far from simplifying the strategies, adds an element of complexity. The new intelligent aids designed to assist the operator in these tasks begins to advise him on what to do, but still remain silent about when to act.

4. Discussion and forms of errors

The operator bases his temporal strategies on the checking of cues, they thus have to be visible. His pinpointing is focused on the critical phases of the system's evolution, where he still has to perform actions. As the technological development has delocated his traditional area of actions, from the process to the whole system, the non-visibility of the team actions increases the complexity of his task. But what kind of knowledge is encompassed within the temporal envelopes? The rhythm of the checking follows the dynamic evolution of the system; it is not exclusively the result of an internal clock. But what really suggests a knowledge is the semantic of the choice of the cues: these are samples of the critical parameters of the system at the very moment of the checking. Which parameters are really critical is a subjective matter: operators and engineers differ in their choice. The operators seem to be more cautious; their checking contains three types of parameters: (i) the critical parameters from a functional point of view; (ii) the unreliable parameters; (iii) the parameters giving the state of the system's elements they do not understand very well. This abundance in the checking reflects their awareness of knowledge and control limitations. If they have schemata of most of the usual situations, and this is what the behavioural traces and the verbalizations suggest, they have a bad envisioning of the future of the system. This is especially the case when notions such as gradients and exponential evolutions are brought into play. However, in intrinsically complex situations, there is no mathematical model to predict the future state of the system, and if these situations frequently occur, the operators finally succeed in finding adequate behavioural responses to control them.

Many studies have tried to demonstrate the superiority of one type of knowledge over another one: the theoretical knowledge of the engineers versus the practical knowledge of the operators. The results of such comparisons have always been ambiguous and spurious. They have neglected to take into account the interactions between the knowledge and the characteristics of the situation; they have ignored the successful efforts developed by the operators to compensate for the gaps in their educational background by the use of checking and redundant mechanisms of time estimation.

But temporal errors exist. They can occur for trivial reasons. Disturbance of basic mechanisms of time estimation, lack of attention, etc. These errors could be assimilated to 'temporal slips' (Reason 1989). More interesting are the systematic errors resulting from a mismatch between the operator's cognitive resources, the displays and the situation. From the body of errors that we began to put together two years ago, four critical situations have been isolated: they are situations where only one mechanism is to be used, because of the nature of the task. If in these situations, there is a lack of cognitive resources or of displays, systematic errors occur.

The current technical evolution accentuates the risk of those mismatches. The redundancy of the temporal mechanisms is not favoured by the displays; the clock-time is omni-present. The intrinsic character of complexity of the industrial environment is underestimated by the designers: mathematical models are supposed to control all the cases. Training is rarely focused on temporal aspects, and the difficulties to dynamically integrate an extended and heterogeneous system, composed of interconnected modules, are entirely obliterated. There is very little effort made in communication networks (Falzon 1989), in intelligent decision supports and in modelling.

These contradictions lead to a stress increase for the operator who must attempt to offset the system's deficiencies. In certain cases, he does not succeed; and if his failure brings about a catastrophe, there is nothing more to do than to invoke 'human error'. This magical term covers all in its shadow: the situation, the displays, the designers, and the public is partially reassured.

References

Amalberti, R. *et al.* 1989 Développement d'aides intelligentes au pilotage: formalisation psychologique et informatique d'un modèle de comportement du pilote de combat engagé en mission de pénétration. *Rapp. CERMA* **49**.

Decortis, F. 1988 Dimension temporelle de l'activité cognitive, lors de démarrages de systèmes complexes. *Travail hum.* **51**, 125–138.

Decortis, F., De Keyser, V., Cacciabue, P. C. & Volta, G. 1989 Temporal dimension of man–machine interaction. In *Human–computer interaction and complex systems* (ed. G. R. S. Weir & J. L. Alty). London: Academic Press.

De Keyser, V. 1987 Structuring of knowledge of operators in continuous processes: case study of a continuous casting plant start-up. In *Human error and new technology* (ed. J. Rasmussen, J. Leplat & K. Duncan). London: John Wiley & Sons.

De Keyser, V., Richelle, M. & Crahay, M. 1989 The nature of human expertise. *Tech. Rep. Pol. Sci. Belg.* (Programme I.A., June 1989).

Falzon, P. 1989 *Ergonomie cognitive du dialogue.* Grenoble: Presses Universitaires de Grenoble.

Hayes-Roth, B. & Hayes-Roth, F. 1979 A cognitive model of planning. *Cognitive Science* **3**, 275–310.

Iosif, G. 1968 La stratégie dans la surveillance des tableaux de commande. I. Quelques facteurs déterminants de caractère objectif. *Revue roum. Sci. Soc.* **12**, 147–163.

Iosif, G. 1969 La stratégie dans la surveillance des tableaux de commande. II. Quelques facteurs déterminants de caractère subjectif. *Revue roum. Sci. Soc.* **13**, 29–41.

Kortland, K. & Kragt, H. 1980 Process alarm as a monitoring tool for the operator. Third International Symposium on Loss Prevention and Safety Promotion in the Process Industries. Basle, September 15–19.

Montangero, J. 1979 La genèse des raisonnements temporels. In *Du temps biologique au temps psychologique* (ed. P. Fraisse). Paris: Presses Universitaires de France.

Pedersen, S. A. 1989 Coping with complexity by abstraction and idealization. Presented at the ESPRIT-MOHAWC Workshop. Riso, Roskilde, February 7–9.

Perrow, C. 1984 *Normal accidents.* New York: Basic Books.

Piaget, J. 1946 *La développement de la notion de temps chez l'enfant.* Paris: Presses Universitaires de France.

Reason, J. 1989 *Human error.* Cambridge University Press.

Valax, M. F. 1985 Cadre temporel et planification des tâches quotidiennes. Ph.D. thesis. Université de Toulouse-le-Mirail.

Van Daele, A. 1988 L'écran de visualisation ou la communication verbale? Analyse comparative de leur utilisation par des opérateurs de salle de contrôle en sidérurgie. *Travail hum.* **51**, 65–79.

Van Daele, A. 1989 Dynamic decision making of control room operators in continuous processes: some field study results. In *Comtemporary ergonomics 1989* (ed. E. D. Megaw). London: Taylor and Francis.

Phil. Trans. R. Soc. Lond. B **327**, 577–583 (1990)
Printed in Great Britain

A lattice theory approach to the structure of mental models

By N. Moray

Departments of Mechanical and Industrial Engineering and of Psychology, University of Illinois at Urbana-Champaign, Illinois 61801, *U.S.A.*

Lattice theory is proposed to provide a formalism for the knowledge base used as a mental model by the operator of a complex system. The ordering relation '$\geqslant$' is interpreted as 'is caused by', and the lattice becomes a representation of the operator's causal hypotheses about the system. A given system can be thought of causally in different ways (purposes, mechanics, physical form, etc.). Each gives rise to a separate lattice. These are related to each other and to an objective description of the structure and function of the physical system by homomorphic mappings. Errors arise when nodes on the mental lattices are not connected in the same way as the physical system lattice; when the latter changes so that the mental lattice no longer provides an accurate map, even as a homomorphism; or when inverse one–to–many mapping gives rise to ambiguities. Some suggestions are made about the design of displays and decision aids to reduce error.

1. Introduction

It is often suggested that operators control complex sytems by forming mental models of the system, and it is their mental model that allows them to control a system too large for all variables to be monitored. They use observations of some subset of variables to decide what action must be taken by using their mental model to predict future states of the entire system. (See, for example, many papers in Sheridan & Johannsen (1976), Rasmussen & Rouse (1981), and Moray (1979). The concept of mental models has gained widespread acceptance, but there is no formalism to represent such models. The main purpose of the present paper is to propose a general formalism for mental models.

2. Mental models of systems operation

The ideas here were inspired by a passage in Ashby (1956) where he defines a model and relates it to lattice theory: 'We can now see much more clearly what is meant by a "model"...(Earlier we saw how) three systems were found to be isomorphic and therefore capable of being used as representations of each other...The model will seldom be *iso*morphic with the biological system: usually it will be a homomorphism of it...The model itself is seldom regarded in all its practical detail: usually it is only some *aspect* of the model which is related to the biological system. Thus what usually happens is that the two systems, biological and model, are so related that a homomorphism of one is isomorphic with a homomorphism of the other...The higher the homomorphisms are on their lattices, the better or more realistic will be the model.' (Ashby 1956, p. 109.) Ashby was discussing the way in which biological systems could be modelled by cybernetic physical machines. In this paper I discuss the inverse: how physical machines are modelled in the mind and brain.

Isomorphism and homomorphism are concepts to do with *mappings* between the elements of sets, and this in turn leads naturally to a discussion of lattice theory. Ashby proposes that any large system with interacting subsystems can be represented as a lattice, and that the elements of the lattice represent the decomposition of the system into (relatively independent) subsystems by a series of many–one mappings (Ashby 1956). If a mental model is a lattice mapping of the properties of the real physical system into a knowledge structure in the mind, what kinds of mappings occur, what is the structure of the lattice, and what implications are there for what information can be known and used by the operator?

3. Abstraction hierarchies

Rasmussen (1986) has suggested that complex systems are understood at several levels which together comprise an abstraction hierarchy. At the lowest level there is extremely detailed knowledge about the physical form of a system (nuts, bolts, switches, physical layout). Above that, the system can be thought of as subsystems made up of collections of physical form components, the level of physical function (pumps, valves and heating circuits). Above that is a level of generic function (heating, temperature control, pressure control, parts handling, assembly and maintenance). Above that is the level of abstract function (mass and energy flows, information flows and productivity). Finally, the system can be thought about in terms of its overall goal or purpose.

He suggests that as people perform different tasks they think about the system at different levels. For example, fault diagnosis usually requires concepts at the levels of generic and then physical function, whereas repair and maintenance require understanding at the level of physical function and physical form. The amount of detail varies enormously at the different levels, and what can be thought about depends on the level at which the person is thinking because there is a different decomposition of the system at the different levels. At the highest levels nearly all detail is lost.

This description of a system in terms of an abstraction hierarchy bears a marked resemblance to Ashby's suggestion that a model is a lattice mapping. We can think of a mental model as a series of lattices, each of which contains a certain kind of information, and which map onto one another as homomorphs (and occasionally, in special cases, as isomorphs).

4. A lattice theory of mental models

For a full discussion of lattice theory see, for example, Birkhoff (1948), Szasz (1963) or Skornjakov (1977). A lattice is a partially ordered set (poset) of elements ordered by the relation '$\geqslant$'. For all the sets with which we are concerned, the set is closed, as physical systems have a finite number of parts. The set has a least element (0) and a greatest element (1). Elements of the set can be represented as points on a diagram. Elements higher on the lattice are composed of unions of elements lower on the diagram. If any two points are joined by a rising line they are related by the order relation $\geqslant$. Only elements so connected are related (no horizontal lines are permitted). A new lattice can be formed from an existing lattice by means of mapping rules. If the mapping is '1–to–1 and onto', then the new lattice is said to be isomorphic with the original lattice. If the mapping is many to one, the mapping is homomorphic. We agree with

Ashby that a lattice formed from a previous lattice by mapping is by definition a *model* of that lattice. In general, inverse mappings can be performed, but will only recover the original lattice for isomorphisms, not homomorphisms. Figure 1 shows examples of graphs that include lattices and examples of mappings.

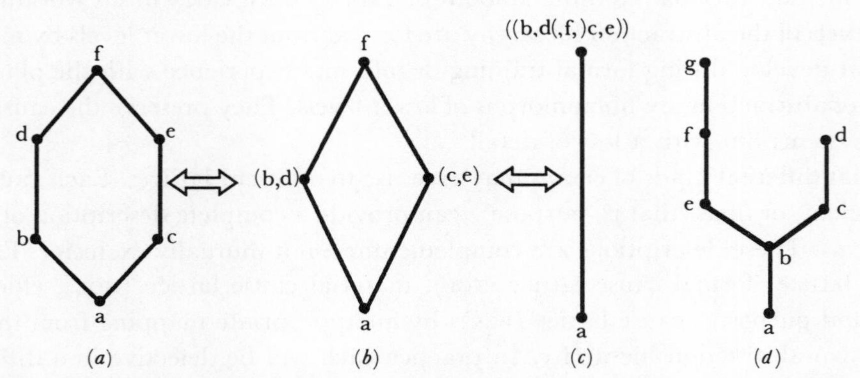

FIGURE 1. Examples of lattices and mappings. Parts (*a*)–(*c*) represent a lattice with progressive simplification by means of homomorphic mapping. The arrows represent mappings, which can be in either direction. Part (*d*) is not a lattice because there is no single greatest element to which the topmost elements are both connected.

The ability of lattice theory to represent mental models of systems knowledge arises when we interpret the order relation $\geqslant$. The algebraic interpretation is 'greater than or equal to', but the notation supports any interpretation that preserves partial ordering. Consider the ways in which a person may think about the functional and causal relations between parts of a system. Classically (in the sense of going back at least to Aristotle), four kinds of cause have been distinguished. A switch may cause a pump to operate because it is the 'on' position (formal cause), because it closes a pair of contacts (material cause), because it allows current to flow through the pump (efficient cause), or because we need cooling (final cause). But causality is an ordering relation in physical systems. A switch causes a pump to operate, which causes coolant to flow, which causes heat removal, which controls the reaction temperature, and so on. We can represent the causal relations between the parts of a system by a lattice, interpreting '$\geqslant$' to mean 'is caused by'. If two elements on the lattice are connected by a line, the lower one is the cause of the upper, and the upper one the effect of the lower.

We start with an objective lattice description of the real physical relations between the parts of the system as described in engineering specifications, or (which ideally we take to be equivalent) as discovered by an exhaustive examination of interactions among the physical components of the real plant. This lattice we will call the physical system lattice (PSL). In so far as an operator's mental model is isomorphic to the PSL, just to that extent is it a complete model of the physical system; and just to that extent will the mental model's predictions exactly match the output of the different parts of the physical system when it is provided with system inputs and parameter values. In general, however, the operator's knowledge will be imperfect for at least two reasons. First, if the system is large, it may simply be impossible for the operator to scan and remember the displayed values of the system variables so as to acquire a perfect knowledge of the system relations. (Indeed some of them may not be displayed.) Second, and perhaps more importantly, the abstraction hierarchy suggests that for many purposes mental models will be homomorphs, not isomorphs of the physical system.

The higher the level of the abstraction hierarchy at which a person thinks about the system, the fewer the elements there are to think about. A 'cooling system' may contain several 'pumps'. A 'pump' may contain several 'glands'. A 'gland' may contain several 'seals'. Thus it is advantageous for an operator to think about a system as high up the hierarchy as possible to reduce his mental workload and the amount of data he must carry in his working memory. The higher levels of the abstraction hierarchy are formed from the lower levels by many-to-one mappings that develop during formal training or informal experience with the plant. That is, higher levels of abstraction are homomorphs of lower levels. They preserve the causal relations between subsystems but with a loss of detail.

Suppose that different kinds of causes may give rise to different lattices. Each cause (formal, material, efficient, or final (that is, purpose)) can provide a complete description of the system in its own terms. These descriptions are complementary, not mutually exclusive. Each can be derived as a lattice (formal cause lattice (FCL), material cause lattice (MCL), efficient cause lattice(ECL) and purposive cause lattice (PCL)) by an appropriate mapping from the PSL; and each has its own abstraction hierarchy. In practice each will be defective in a different way. For example, one may know that a particular circuit is present to provide cooling (final cause) and know what values of the display show that it is working and what controls switch it on or off (formal cause), but not know what mechanism is involved or its underlying physical principles (material and efficient cause). In such a case the FCL and PCL lattices will contain elements not present in the posets of the MCL and ECLS. Other examples will occur to the reader.

5. Mental models as sets of lattices

We see therefore that several kinds of knowledge (the causal lattices) can be obtained from knowledge about the original physical system. These causal lattices are derived from the PSL by homomorphic mappings. To complete the picture we assume that they can be mapped into one another. That is, they are homomorphs of one another. This allows operators to change the way in which they think about the problem as well as the level at which they think about the system. Some of the relations among the elements of the mental model are shown in figure 2.

We cannot here consider how learning mechanisms lead to the construction of the lattices and the original mappings. Those features of the theory have yet to be fully worked out, although there seem to be no difficulties in using what is known of induction, concept formation, probability learning, etc. to account for them. But if we examine a steady state of the mental model we can see both the virtues of such a model and also how such a model might lead to error and the inability to manage a faulty system.

Suppose the system operator observes a signal which shows that some component is in an unusual state. This is equivalent to entering one of the lattices at a particular node. Assume that the operator is currently thinking about the component in terms of its purpose. The response to noticing the unusual state will be to ask which other component has as its purpose the control of this component? That question can be answered by going down the PCL lattice from the component in question to the next lower elements on the lattice. If one of them is itself giving an unusual reading, the process is repeated. At some level there will be no unusual component, and the operator will conclude that the lowest greater element above this which has been found abnormal is the 'cause' of the problem. To manage the fault, he must now move to the MCL or ECL lattice, to understand the nature of the physical abnormality that underlies the

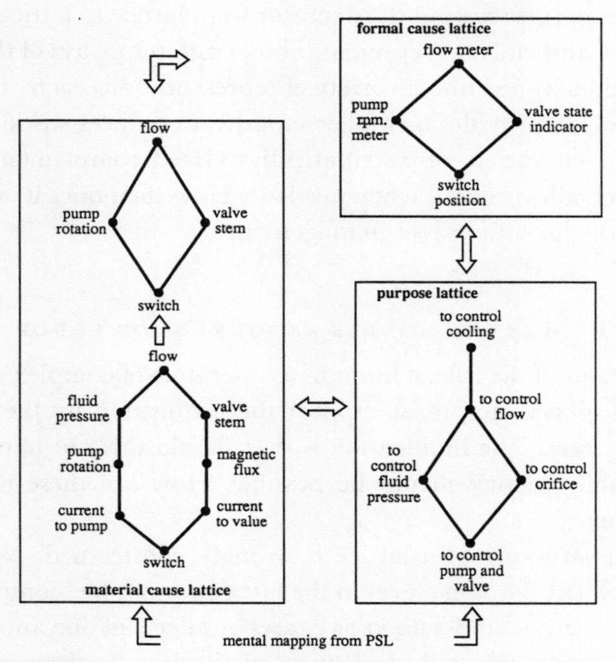

FIGURE 2. Examples of causal lattices and their simplification and relation by means of mappings. The physical system lattice (PSL) is not shown, but represents part of a cooling system operated by a double pole switch that operates both a pump and an electromagnetic valve to supply cooling water. Note that the existence of mappings does not guarantee that an equivalent node will be available when moving from one lattice to another.

abnormal reading. A mapping carries him into the appropriate lattice at the level where the abnormal PCL was identified, and on this new lattice he once again travels up and down the lattice seeking the lowest level at which there is an abnormal element. That element is the appropriate explanation for the fault. By travelling along the paths of the lattice, and by moving from lattice to lattice, a number of strategies may become apparent for managing the fault. It may be that on returning to the original lattice at its original entry point it is possible to find another component below the one which was abnormal, which is inactive, but which has the same purpose. By then mapping back onto the FCL, the way to switch on that component will be found. The result will be that the operator turns on a redundant standby component and the fault is brought under control.

The multiple representation of knowledge allows the operator to think in a new way about a problem. If he travels all the paths of a particular lattice and does not find any relevant information for the task in hand, he will be forced to try a different lattice. For example, if as a result of an alarm he checks all controls (FCL) and cannot find any abnormal settings, he will have to think about other reasons for the alarm being on, which will suggest a move to one of the other causal lattices. On the other hand, suppose that following the alarm he asks himself a series of questions about the component such as, 'What is it for? What turns it on? Where is its value displayed? What is the underlying physics?' he is staying at the same level of the abstraction hierarchy (same height on the lattice) but moving from lattice to lattice. If no satisfactory information is forthcoming, he will have to change the level of abstraction and try again. In this case exhausting a particular strategy forces him to change levels in each lattice rather than changing from lattice to lattice.

Given a set of switching rules to send the operator from lattice to lattice, such a mental model provides an economical and efficient representation of different ways of thinking about the real system: the PSL has been mapped into a variety of representations each of which provides many levels of abstraction (which provides mental economy and reduces workload if the appropriate level is chosen) and a rich variety of systematically related complementary ways of thinking about a problem (which allows detail when needed). How then does it account for error, what suggestions does it make for aiding risk management?

6. Mental models as sources of error

In almost all discussions of the role of humans as operators of complex systems statements are made that training, displays, etc., must support the acquisition by the operator of a correct mental model of the system. The implication is that should this not happen, errors will occur and efficient fault management will not be possible. How are these claims reflected in the properties of lattice theory?

First, recall that the structure of a lattice is strongly constrained. No horizontal links are possible. Hence variables at the same level in the lattice cannot be 'compared'. Only variables that are related by '$\geqslant$' can be understood as causes or effects of one another. If in the original learning mistakes are made about the structure of the lattice, these mistakes may make it impossible to understand certain components as linked in cause–effect relations. Those relations are literally unthinkable, as only lattice comparable elements have a conceivable relationship.

Secondly, one lattice is usually derived from another by homomorphic rather than isomorphic mapping. This has the advantage of reducing workload for normal situations, but means that the inverse mapping is one–to–many, which means in turn that there is bound to be uncertainty in choosing a route downwards on the lattice. Upward homomorphism reduces 'variety' in Ashby's phrase, and purchases economy of effort at the price of loss of Requisite Variety (Ashby 1956). Hence, there is a possibility that the pursuit of efficiency in the operation of normal systems will be effected at the cost of ambiguity when handling abnormal situations. Note that whenever we construct a physical system that is so complex that it forces the operator to make use of lattice homomorphism to handle it we guarantee that this problem will exist.

Thirdly, there is only a homomorphism, not an isomorphism, between the four kinds of causal lattice. It may be that a person has no structure in the lattice that represents formal relations (FCL) and corresponds to the way in which he or she thinks about the purposes of subsystems (PCL). Things that can be thought about one way cannot then be thought about in another easily. But the latter may be the only way, for a given PSL, to solve the problem. The constraints imposed by lattice structure and homomorphic mapping has the effect of literally making certain things unthinkable, and the more solidly the model becomes entrenched, the less is it possible to think those thoughts. Only relations that are represented on at least one lattice, can be thought about at all, and only those represented on several can be thought about in different ways. There are rigorous ways of examining the PSL and predicting which relations are the natural ones to be embodied in the mental lattices, but they cannot be considered here (for example, see Conant (1976)).

Finally, because of the ordering requirements of a lattice, closed loop structures cause a particular difficulty, because they contain components that are both above and below each other on the lattice, and that is prohibited. I have suggested elsewhere how this might be represented in a lattice framework, and the expected psychological consequences for people trying to diagnose and manage faults containing closed loops (Moray 1988). Briefly, lattice theory predicts that they will either treat the entire closed loop system as an undecomposable black box, or they will try to think about it piecewise as a series of open loop components. There is some empirical evidence for the latter.

What do these results tell us about design for decision aids? Displays and training should indeed help an operator to construct an accurate mental model of the systems with which he or she must work. But they must also be designed so as to make accessible those relations in the system that are likely to be omitted from the causal lattices. This requires an analysis of the PSL to determine its natural decomposability into subsystems (Conant 1976). It is those that are most likely to be preserved in the mental model. In faulty systems couplings change, and displays and decision aids should provide ways to display alternative couplings that are least likely to be built into the mental model, and enable to support the operator when totally new structures emerge. So despite the value of the stable mental model, it must be modified if the PSL changes, although this is a slow process involving learning, whereas movements from lattice to lattice or along paths in a lattice is rapid and voluntary. The systematic analysis of PSL details is a strong strategy for predicting what kinds of faults are most likely to cause severe difficulties for operators when they must perform fault management.

7. Conclusions

There is at present no formalism for describing the structure and content of mental models, although it is widely held that they play a central role in the interaction between humans and complex systems that they operate. The examples given in this paper suggest that lattice theory may be a suitable formalism, and that the properties of posets and homomorphic mappings may provide a representation of how different kinds of causal knowledge are inter-related in mental models, how certain kinds of errors arise, and steps that may help operators in times of emergencies. As formalisms provide strong predictions for empirical tests, this approach may help to move us from purely qualitative accounts of mental models and error to a more powerful basis for empirical studies, as well as providing a basis for the rational design of aids to reduce human error and its impact.

This work was supported by a grant from the University of Illinois Research Board which is gratefully acknowledged.

References

Ashby, W. R. 1956 *Introduction to cybernetics*. London: Chapman and Hall.
Birkhoff, G. 1948 Lattice theory. *Am. math. Soc.* (collected publications), vol. XXV.
Conant, R. C. 1976 Laws of information which govern systems. *IEEE Trans. Syst. Man. and Cyber.* **6**, 240–255.
Moray, N. 1979 *Mental workload.* London and New York: Plenum Press.
Moray, N. 1988 A lattice theory of mental models of complex systems. *Tech. Rep. EPRL-88-08* (Engineering Psychology Research Laboratory, University of Illinois).
Rasmussen, J. 1986 *Information processing and human machine interaction.* Amsterdam: North-Holland Press.
Rasmussen, J. & Rouse, W. B. 1981 *Human detection and diagnosis of system failures.* New York: Plenum Press.
Sheridan, T. B. & Johannsen, G. 1978 *Monitoring behavior and supervisory control.* New York: Plenum Press.
Skornjakov, L. A. 1977 *Elements of lattice theory.* Delhi: Hindustan Publishing (India); Bristol: Adam Hilger Ltd.
Szasz, G. 1963 *Introduction to lattice theory.* New York: Academic Press.

Phil. Trans. R. Soc. Lond. B **327**, 585–593 (1990)
Printed in Great Britain

The 'problem' with automation: inappropriate feedback and interaction, not 'over-automation'

By D. A. Norman

Department of Cognitive Science, University of California, San Diego, California 92093, *U.S.A.*

As automation increasingly takes its place in industry, especially high risk industry, it is often blamed for causing harm and increasing the chance of human error when failures do occur. I propose that the problem is not the presence of automation, but rather its inappropriate design. The problem is that the operations under normal operating conditions are performed appropriately, but there is inadequate feedback and interaction with the humans who must control the overall conduct of the task. When the situations exceed the capabilities of the automatic equipment, then the inadequate feedback leads to difficulties for the human controllers.

The problem, I suggest, is that the automation is at an intermediate level of intelligence, powerful enough to take over control that used to be done by people, but not powerful enough to handle all abnormalities. Moreover, its level of intelligence is insufficient to provide the continual, appropriate feedback that occurs naturally among human operators. This is the source of the current difficulties. To solve this problem, the automation should either be made less intelligent or more so, but the current level is quite inappropriate.

The overall message is that it is possible to reduce error through appropriate design considerations. Appropriate design should assume the existence of error, it should continually provide feedback, it should continually interact with operators in an effective manner, and it should allow for the worst situations possible. What is needed is a soft, compliant technology, not a rigid, formal one.

The 'problem' with automation: inappropriate feedback and interaction, not 'over-automation'

Although automation is often identified as a major culprit in industrial accidents, I propose that the problems result from inappropriate application, not the commonly blamed culprit, 'over-automation'. According to this view, operations would be improved either with a more appropriate form of automation or by removing some existing automation. Current automatic systems have an intermediate level of intelligence that tends to maximize difficulties.

This leads to a second point, namely, that in design, it is essential to examine the entire system: the equipment, the crew, the social structure, learning and training, cooperative activity, and the overall goals of the task. Analyses and remedies that look at isolated segments are likely to lead to local, isolated improvements, but they may also create new problems and difficulties at the system level. Too often, the implementation of some new 'improved' automatic system, warning signal, re-training, or procedure is really a sign of poor overall design: had the proper system level analysis been performed, quite a different solution might have resulted.

Automation: simultaneously too much and too little

Consider the task of the crew on a modern commercial airplane. Most of the flight activity is routine. Large, modern aircraft are relatively easy to fly: the airplane is stable, responsive, and manoeuverable. The automatic equipment monitors all operations and helps ease the workload of the crew. Indeed, whereas the commercial airplane of a few years ago required a crew of three, the newer planes need only two people to fly them, and most of the time, only one is really necessary. Most of this is good, and the accident rate with modern aircraft has been decreasing over the years, the decrease highly correlated with (and usually thought to be a result of) the introduction of high technology controls and automation.

There are problems, however. For one, the sheer size of the plane means that the crew cannot know all that is happening. They are physically isolated from the passengers and from any difficulties that may be occurring within the passenger section of the plane. They are isolated from most of the physical structures of the aircraft. Even more important than physical isolation is the mental isolation caused by the nature of the controls. The automation tends to isolate the crew from the operations of the aircraft because the automatic equipment monitors and controls the aircraft, providing little or no trace of its operations to the crew, isolating them from the moment-to-moment activities of the aircraft and of the controls. On the one hand, this combination of relative physical and mental isolation from the details of flying helps contribute to the safety by reducing workload and reliance on possible human variability or failure. On the other hand, when the automatic equipment fails, the crew's relative isolation can dramatically increase the difficulties and the magnitude of the problem faced in diagnosing the situation and determining the appropriate course of action.

Physical isolation would be all right if the crew were still up to date on the critical states of the device being controlled. The problem is that, increasingly, the physical isolation is accompanied by a form of mental isolation. Zuboff (1989) describes the control room of a modern paper mill: where once the operators roamed the floor, smelling, hearing and feeling the processes, now they are poised above the floor, isolated in a sound-isolated, air-conditioned, glass control room. The paper-mill operators do not get the same information about the state of the mill from their meters and displays as they did before from the physical presence. The ship captain does not have a good feel for the actual problems taking place on the other side of the ship. And the automatic equipment in an airplane cockpit can isolate the crew from the state of the aircraft. It is this mental isolation that is thought to be largely responsible for many of the current difficulties.

Detecting system problems: three case studies

Here are three case studies from the world of aviation, the domain chosen because aviation is the best documented and validated of all industrial situations.

(a) *The case of the loss of engine power*

In 1985, a China Airlines 747 suffered a slow loss of power from its outer right engine. This would have caused the plane to yaw to the right, but the autopilot compensated, until it finally reached the limit of its compensatory abilities and could no longer keep the plane stable. At that point, the crew did not have enough time to determine the cause of the problem and to take action: the plane rolled and went into a vertical dive of 31500 feet before it could be recovered. The aircraft was severely damaged and recovery was much in doubt (NTSB 1986; Wiener 1988).

(b) *The case of the 'incapacitated' pilot*

The second case study is a demonstration that lack of information and interaction can take place even in the absence of automation, an important piece of evidence for my argument that automation *per se* is not the key issue.

In 1979, a commuter aircraft crashed while landing at an airport on Cape Cod, Massachusetts (U.S.A.), killing the captain and seriously injuring the first officer and six passengers. The first officer observed that the approach was too low and commented on this to the captain. However, the captain did not respond. But the captain, who was also president of the airline, and who had just hired the first officer, hardly ever responded, even though airline regulations require pilots to do so. Moreover, the captain often flew low. There were obvious social pressures on the first officer that would inhibit further action.

What the first officer failed to notice was that the captain was 'incapacitated', possibly even dead from a heart attack. The U.S. National Transportation Safety Board (NTSB) described it this way: 'The first officer testified that he made all the required callouts except the "no contact" call and that the captain did not acknowledge any of his calls. Because the captain rarely acknowledged calls, even calls such as one dot low (about 50 ft below the 3° glide slope) this lack of response probably would not have alerted the first officer to any physiologic incapacitation of the captain. However, the first officer should have been concerned by the aircraft's steep glidepath, the excessive descent rate, and the high airspeed' (NTSB 1980).

Before you think this a strange, isolated instance, consider this. In an attempt to understand this rather peculiar circumstance, the NTSB noted that United Airlines had earlier performed a study of simulated pilot incapacitation: 'In the United simulator study, when the captain feigned subtle incapacitation while flying the aircraft during an approach, 25 percent of the aircraft hit the "ground". The study also showed a significant reluctance of the first officer to take control of the aircraft. It required between 30 sec and 4 min for the other crewmember to recognize that the captain was incapacitated and to correct the situation' (NTSB 1980).

(c) *The case of the fuel leak*

In the previous two case studies, the crew was unaware of the developing problems. In this third case study, the vigilant second officer noticed one sign of a problem, but failed to detect another. Here is a quotation from the accident report filed with the NASA Aviation Safety Reporting System (Data Report 64441, February 1987) (these are voluntary reports, submitted by the people involved). 'Shortly after level off at 35,000 ft.... the second officer brought to my attention that he was feeding fuel to all 3 engines from the number 2 tank, but was showing a drop in the number 3 tank. I sent the second officer to the cabin to check that side from the window. While he was gone, I noticed that the wheel was cocked to the right and told the first officer who was flying the plane to take the autopilot off and check. When the autopilot was disengaged, the aircraft showed a roll tendency confirming that we actually had an out of balance condition. The second officer returned and said we were losing a large amount of fuel with a swirl pattern of fuel running about mid-wing to the tip, as well as a vapor pattern covering the entire portion of the wing from mid-wing to the fuselage. At this point we were about 2000 lbs. out of balance....'

In this example, the second officer (the flight engineer) provided the valuable feedback that something seemed wrong with the fuel balance. The automatic pilot had quietly and efficiently compensated for the resulting weight imbalance, and had the second officer not noted the fuel discrepancy, the situation would not have been noted until much later, perhaps too late.

Suppose the automatic pilot could have signalled to the crew that it was starting to compensate the balance more than was usual, or at the least, more than when the autopilot was first engaged? This would have alerted the crew to a potential problem. Technically, this information was available to the crew, because the autopilot controls the aircraft by physically moving the real instruments and controls, in this situation, by rotating the control wheel to maintain balance. The slow but consistent turning of the wheel could have been noted by any of the three crew members. This is a subtle cue, however, and it was not noted by either the pilot or the co-pilot (the first officer) until after the second officer had reported the fuel unbalance and had left the cockpit.

The problem is not automation, it is lack of feedback

Automation is increasingly blamed for problems in high-risk industry. The general theme of the argument is that in the 'good old days', before automation, the controllers were actively engaged in the plant operation. They had to monitor everything and control everything. This had problems, in particular high mental workloads and over-reliance on people's abilities to be continually alert, accurate, and knowledgeable. But it had the virtue of keeping the operators continually informed as to the state of the system.

In the language of control theory or servomechanisms, a system has a desired state, a means for adjusting the system toward that desired state, and then a feedback loop in which the actual state of the system is compared with the desired state, so that additional correction can be performed if there is a mismatch. The combination of this control plus feedback is called the control loop, and when a human is operating the equipment manually, the human is an essential element of the control loop: hence the saying, 'the person is in the loop'. With the advent of automated controls, the human's job changed. The automation took care of the lower level actions and the human operators simply watched over the system, presumably ever-alert for deviations and problems. Now the human operators were managers or supervisors rather than controllers: they were 'out of the loop' (see the papers in Rasmussen & Rouse (1981), Bainbridge (1987), Norman & Orlady (1989), or Weiner & Currey (1980)).

Automation has clearly improved many aspects of performance. It leads to superior productivity, efficiency, and quality control. In aircraft, fuel efficiency is improved, schedules can be maintained in inclement weather, and overall accident rates have gone down. But automation also leads to difficulties. When problems arise, the crew may not be sufficiently up-to-date with the current state of the system to diagnose them in reasonable time, and the general reliance on automatic systems may have led to a degradation of manual skills. Finally, the highest stress and workload occur at times of trouble. That is, automatic equipment seems to function best when the workload is light and the task routine: when the task requires assistance, when the workload is highest, this is when the automatic equipment is of least assistance: this is the 'irony' of automation (Bainbridge 1987; see also Norman & Orlady (1989)).

What of the fact that the people are 'out of the loop'? Is this the major culprit? In some of the case studies in this paper, the crew was clearly out of the loop, failing to detect symptoms of trouble early enough to do anything about them. But in one of the studies, the case of the 'incapacitated' pilot, no automation was involved. Instead, there was an uncommunicative captain, plus social pressures that worked against a junior first officer interrupting the activities of a senior captain. In other words, although the human operators are indeed no longer 'in the loop', the culprit is not automation, it is the lack of continual feedback and interaction.

Two thought experiments

Consider two thought experiments. In the first, imagine a captain of a plane who turns control over to the autopilot, as in the case studies of the loss of engine power and the fuel leak. In the second thought experiment, imagine that the captain turns control over to the first officer, who flies the plane 'by hand'. In both of these situations, as far as the captain is concerned, the control has been automated: by an autopilot in one situation and by the first officer in the other. But in the first situation, if problems occur, the autopilot will compensate and the crew will notice only by chance (as in the case study of the fuel leak). When automatic devices compensate for problems silently and efficiently, the crew is 'out of the loop', so that when failure of the compensatory equipment finally occurs, they are not in any position to respond immediately and appropriately.

In the case of the second-thought experiment where the control was turned over to the first officer, we would expect the first officer to be in continual interaction with the captain. Consider how this would have worked in the case studies of the loss of engine power or the fuel leak. In either case, the problem would almost definitely have been detected much earlier in the flight. The first officer would probably have said something like 'I seem to be correcting this thing more and more, I wonder what's happening?' Yes, from the captain's point of view the rest of the crew serves as a type of automaton, but one that observes and remarks upon conditions. By reporting upon observations and possible discrepancies, each crew member keeps the rest informed and alerted: keeping everyone 'in the loop'.

The observations of these thought experiments are buttressed by the situation described in the case study of the fuel leak, where the second officer, routinely scanning the gauges, noted a puzzling discrepancy and commented on it to the captain. As the captain's report said, 'the second officer brought to my attention that he was feeding fuel to all 3 engines from the number 2 tank, but was showing a drop in the number 3 tank. I sent the second officer to the cabin to check that side from the window'. Here, even though the second officer did not understand the reason for the discrepant fuel gauge reading, the voiced observation prompted the captain to look over the aircraft by sending the second officer to the cabin to examine the wing and for himself to check the cockpit. The cockpit check led the captain to note that the 'wheel was cocked to the right', which then led to the discovery of the weight imbalance caused by a massive fuel leak. At the time the second officer commented on the fuel gauge reading, he did not know what the problem was, but his comment alerted the crew.

Again, this observation makes the point that the culprit is not actually automation, but rather the lack of feedback. The informal chatter that normally accompanies an experienced, socialized crew tends to keep everyone informed of the complete state of the system, allowing for the early detection of anomalies. Hutchins (1990) has shown how this continual verbal interaction in a system with highly social crews serves to keep everyone attentive and informed, helps the continual training of new members of the crew, and serves as natural monitors for error.

The solution? More appropriate automation

The message is that automation, *per se*, is not the culprit in high-risk situations. Many of the current problems are indeed a result of automation, but only in the sense that the automation is inappropriately designed and applied.

When people perform actions, feedback is essential for the appropriate monitoring of those actions, to allow for the detection and correction of errors, and to keep alert. This is hardly a novel point: feedback is an essential aspect of all control theory. But adequate feedback to the human operators is absent far more than it is present, whether the system be a computer operating system, an autopilot, or a telephone system. In fact, it is rather amazing how such an essential source of information could be skipped: the need for complete feedback is one of the major points of Norman (1988). Without appropriate feedback, people are indeed out of the loop: they may not know if their requests have been received, if the actions are being performed properly, or if problems are occurring. Feedback is also essential for learning, both of tasks, and also of the way that the system responds to the wide variety of situations it will encounter.

People construct mental models of systems with which they interact. The model is constructed entirely from what I have called 'the system image', the information available to them from the system, the environment, and their instructions (Norman 1986). But this system image depends critically upon the information displays of modern equipment. When we send a command to an automated piece of equipment, the only way we can update our mental models of the system is through the feedback provided to us.

In the first case study, the China Airlines situation where the autopilot kept compensating for the loss of engine power, if the autopilot had been intelligent enough, it might have reported the need to keep compensating. In the case study of the weight imbalance caused by a fuel leak, there were two opportunities to note the problem. An intelligent automaton could have reported on the continual increase in compensation necessary to keep the plane level. Or it might have noted that the fuel level of the number three tank was falling, even though fuel was only supposed to be pumped from the number two tank. And in the case of the incapacitated pilot, if the captain and his first officer had been better socialized and had followed normal and proper callout and response procedures with the two considered as equal members of the operation, the pilot's incapacitation would have been discovered.

(*a*) *We do not know enough to mimic natural human interaction*

Note that the problems in all three of the case studies were not because of lack of information, at least not in the technical sense. Autopilots work by physically moving the same controls that the pilots use. In the case studies of the loss of engine power and the fuel leak, the autopilots compensated by turning the control wheels. In theory, the crew could have noted the problem quite early by noting the position of the wheels, just as the second officer noted an abnormality in the fuel gauge readings in the fuel leak case study. Similarly, there was sufficient information in the case of pilot incapacitation. In these cases the problem was that no person or system commented upon the issues, so that nothing brought the potential problem to the attention of the relevant people. The feedback was potentially available, but it was not attended to properly.

During the writing of this paper, I took part in an informal replication of the fuel leak incident in the NASA-Ames full-vision, full-motion 727 simulator. Once again, the second officer failed to note the discrepant control wheel position, even though in this case he had read the relevant accident report: the normal cockpit activities drew the focus of attention away from the control wheel position. Our analyses afterwards showed that the wheel position was not a very salient clue in any case. We plan further studies including a careful replication of this situation as well as a formal experimental study of the two 'thought experiments' described in this paper.

The task of presenting feedback in an appropriate way is not easy to do. Indeed, we do not yet know how to do it. We do have a good example of how not to inform people of possible difficulties: overuse of alarms. One of the problems of modern automation is the unintelligent use of alarms, each individual instrument having a single threshold condition that it uses to sound a buzzer or flash a message to the operator, warning of problems. The proliferation of these alarms and the general unreliability of these single-threshold events causes much difficulty (see Patterson, this symposium; Sorkin (1989), and Sorkin *et al.* (1988)). What is needed is continual feedback about the state of the system, in a normal natural way, much in the manner that human participants in a joint problem-solving activity will discuss the issues among themselves. This means designing systems that are informative, yet non-intrusive, so the interactions are done normally and continually, where the amount and form of feedback adapts to the interactive style of the participants and the nature of the problem. We do not yet know how to do this with automatic devices: current attempts tend to irritate as much as they inform, either failing to present enough information or presenting so much that it becomes an irritant: a nagging, 'back-seat driver', second-guessing all actions.

(*b*) *A higher order of awareness is needed*

To give the appropriate kind of feedback requires a higher level of sophistication in automation than currently exists. Consider what is required for an automatic pilot to note that it is compensating more than normal. The current automatic systems are feedback loops that attempt to maintain a constant system state. To provide self-monitoring capability that would let it recognize that conditions are changing and more and more compensation is being used, would require a kind of higher level of awareness, a monitoring of its own monitoring abilities.

Now, obviously, it would not be difficult to build automatic systems for the specific cases of monitoring for increased rudder or control-yoke compensation, or for inappropriate fuel loss: any competent computer-scientist could write an appropriate program. But what about the next problem, one that will involve yet a different system, yet a slightly different anomaly? We do not know how to solve the general condition.

Consider what would be required of a fuel monitoring system to detect that the fuel level of tank *x* was dropping, but that fuel was only supposed to be fed from tank *y*. To solve this problem, in the general case, requires an intelligent system, one that understands the implications of the various control settings of the system. There probably has to be a knowledge base of the systems in the aircraft plus an internal representation for the items that would allow the system to reason about the potential cases. This is the sort of thing done today in laboratories of artificial intelligence and cognitive science, but we do not know how to solve this problem, for the general case. Moreover, even if the automatic monitoring equipment were to note the existence of a system trend or discrepancy that could lead to a difficulty later on, how should it be brought to the attention of the operators in a natural, intelligent fashion, much the way that normal cockpit conversation works?

The solutions will require higher levels of automation, some forms of intelligence in the controls, an appreciation for the proper form of human communication that keeps people well informed, on top of the issues, but not annoyed and irritated. Our current level of knowledge is not enough to do these things.

(*c*) *The new irony of over-automation*

Many ills have been laid at the feet of 'over-automation'. Too much automation takes the human out of the control loop, it deskills them, and it lowers morale. One much remarked-upon irony of automation is that it fails when it is most needed. I agree with all the analyses of the problems, but from these analyses, I reach the opposite conclusion, a different irony: our current problems with automation, problems that tend to be blamed on 'over-automation', are probably the result of just the opposite problem: the problem is not that the automation is too powerful, the problem is that it is not powerful enough.

(*d*) *Why don't current systems provide feedback?*

Why do current systems have such poor feedback and interaction? In part, the reason is a lack of sensitivity on the part of the designer, but in part, it is for a perfectly natural reason: the automation itself doesn't need it! That is, if a designer is asked to design an automatic piece of equipment to control some function, the task is completed when the device functions as requested. Providing feedback and monitoring information to the human operators is of secondary importance, primarily because there does not appear to be any need for it.

Feedback is essential because equipment does fail and because unexpected events do arise. In fact, in any complex tasks or environment, one should always expect unexpected events: what is unexpected is the type of event that will occur. Human operators need to cope with these situations, and this is why the feedback and 'conversation' is required. Were the equipment never to fail, were it capable of handling all possible situations, then the human operator would not be necessary, so the feedback and interaction would similarly not be necessary. Today, in the absence of perfect automation an appropriate design should assume the existence of error, it should continually provide feedback, it should continually interact with operators in an appropriate manner, and it should have a design appropriate for the worst of situations. What is needed is a soft, compliant technology, not a rigid, formal one.

This research was supported by grant NCC 2-591 to Donald Norman and Edwin Hutchins from the Ames Research Center of the National Aeronautics and Space Agency in the Aviation Safety/Automation Program. Everett Palmer served as technical monitor. Additional support was provided by funds from the Apple Computer Company and the Digital Equipment Corporation to the Affiliates of Cognitive Science at UCSD. Earlier research has been supported by a grant from the Nippon Telegraph and Telephone Company and the System Development Foundation. I thank Edwin Hutchins for his interactions and guidance and Jim Hollan, Julie Norman, Hank Strub and Nick Flor for their comments on the manuscript.

References

Bainbridge, L. 1987 Ironies of automation. In *New technology and human error* (ed. J. Rasmussen, K. Duncan & J. Leplat). New York: Wiley.

Hutchins, E. 1990 The technology of team navigation. In *Intellectual teamwork: the social and technological foundation of collaborative work* (ed. J. Galegher, R. Kraut & C. Ejido). New Jersey: Lawrence Erlbaum Associates.

Norman, D. A. 1986 Cognitive engineering. In *User centered system design* (ed. D. A. Norman & S. W. Draper). New Jersey: Lawrence Erlbaum Associates.

Norman, D. A. 1988 *The psychology of everyday things*. New York: Basic Books.

Norman, S. D. & Orlady, H. W. (eds) 1989 *Flight deck automation: promises and realities*. Moffett Field, California: National Aviation and Space Administration.

NTSB (National Transportation Safety Board) 1980 *Aircraft accident report – Air New England, Inc. DeHavilland DHC-6-300, N383 EX Hyannis, Massachusetts. June 17, 1979* (*Report No. NTSB/AAR-80/01*.) Washington: NTSB.

NTSB (National Transportation Safety Board) 1986 *Aircraft accident report – China Airlines 747-SP, N4522V, 300 Nautical Miles Northwest of San Francisco, California, February 19, 1985*. (*Report No. NTSB/AAR-86/03*.) Washington: NTSB.

NTSB (National Transportation Safety Board) 1988 *Aircraft accident report – Northwest Airlines, Inc., McDonnell Doublas DC-9-82, N312RC, Detroit Metropolitan Wayne County Airport, Romulus, Michigan, August 16, 1987*. (*Report No. NTSB/AAR-88/05, Govt. Accession No. PB 88-91046. May 10, 1988*.) Washington: NTSB.

Rasmussen, J. & Rouse, W. B. (eds) 1981 *Human detection and diagnosis of system failures*. New York: Plenum Press.

Sorkin, R. D. 1989 Why are people turning off our alarms? *Hum. Fact. Soc. Bull.* **32**, 3–4.

Sorkin, R. D., Kantowitz, B. K. & Kantowitz, S. C. 1988 Likelihood alarm displays. *Hum. Fact.* **30**, 445–459.

Weiner, E. L. 1988 Cockpit automation. In *Human factors in aviation* (ed. E. L. Weiner & D. C. Nagel). San Diego: Academic Press.

Weiner, E. L. & Curry, R. E. 1980 Flight-deck automation: Promises and Problems. *Ergonomics* **23**, 995–1011. Also in *Pilot error: the human factor* (ed. R. Hurst & L. R. Hurst) 1982. New York: Jason Aronson.

Zuboff, S. 1988 *In the age of the smart machine: the future of work and power*. New York: Basic Books.